PRAISE
Coo

"Both literate and earthy, he's not here to convert, proud though he is of his little section of the world. The timing of *Coop* does strike, however unintentionally, at the humbled state of our society. Eggs from backyard chickens are the new pesto. And Perry is an organic granola bar dipped in mud. Wrapped in locally produced bacon."

—Janet Okoben, *Cleveland Plain Dealer*

"Written in engaging prose. . . . The author's keen and often amusing observations about farm chores and the creatures under his care will seem pleasingly familiar to anyone who has spent time on a farm. I often found myself smiling or chuckling. . . . An enjoyable visit to a rural homestead alongside an amiable and benignly quirky companion. . . . Pleasurable."

—Nicolette Hahn Niman, *San Francisco Chronicle*

"Luminous. . . . Perry (*Population: 485*) is that nowadays rare memoirist whose eccentric upbringing inspires him to humor and sympathetic insight . . . . Perry writes vividly about rural life; peck at any sentence . . . and you'll find a poetic evocation of barnyard grace."   —*Publishers Weekly* (starred review)

"You can read Michael Perry's *Coop* as an outrageously funny comedy about a semi-hapless neophyte navigating the pitfalls (and pratfalls) of the farming life. Please do, in fact. But scratch a little deeper, past Perry's lusciously entertaining and epigrammatic prose, his ultra-charming combo of midwestern earnestness and serrated wit, and you'll find a reflective, sincere, and surprisingly touching—at times, even heart-cracking—story about a man struggling to put down roots."
—Jonathan Miles, author of *Dear American Airlines*

"Displays Perry's charming penchant for nurturing things to life—be it a truck or a garden, a community or a baby—while, at the same time, nodding to the past. . . . Perry can take comfort in the power of his writing, his ability to pull readers from all corners onto his Wisconsin spread, and make them feel right at home among the chickens."
—Nicole Brodeur, *Seattle Times*

"Throughout the book . . . I found lovely passages and insights. . . . My vocabulary is richer now with Perry's coined words. . . . You'll find in this book a slender silver cord of smart contemplation about meaning and purpose."
—Susan Ager, *Minneapolis Star Tribune*

"Dryly humorous, mildly neurotic, and just plain soulful—a book that might even make you want to buy a few chickens."
—*Kirkus Reviews* (starred review)

"A new Michael Perry book! *Coop* made me want to (1.) get chickens immediately; (2.) always eat popcorn on Sundays with a lot of people who love one another and who also love popcorn; (3.) birth a baby into a blue tub full of water; and (4.), the easy part, treasure the Wisconsin writer who is funny, lyrical, wise, tender, and sometimes all at once."

—Jane Hamilton, author of

Laura Rider's Masterpiece and *A Map of the World*

"In less talented hands, the stories he recounts in *Coop* would merely have been the subject of an unusually busy holiday letter. But in Perry's engrossing narration, they take on the heft of history, dotted with rueful humor and stories that beg to be performed aloud. . . . His specific experiences are not universal, but the lessons he draws from them are. . . . When tragedy hits him without warning, his perceptions guide us through feelings many will experience but few can articulate with such clarity."   —Rebekah Denn, *Christian Science Monitor*

"Michael Perry delivers when it comes to small town Americana. . . . A perfect Father's Day gift for anyone who's ever tilled the earth; Perry shines as he reminisces. . . . His voice makes *Coop* a heartwarming read chock-full of family values."   —Chris Stuckenschneider, *Missourian*

"I don't know when I've enjoyed a book about country living as much as Mike Perry's *Coop*. As an adventure narrative, Perry's details of rural life are much more telling and authentic than any how-to book. But *Coop* will also appeal to anyone who enjoys good writing. There is a literary gem on nearly every page."   —Gene Logsdon, author of *The Contrary Farmer*

"*Coop*—the title refers to the author's dream project, a chicken coop he builds with his own hands—is typical Perry: written in an easygoing, talk-to-the-ready style, with a self-effacing sense of humor and an ability to conjure up vivid mental pictures with a few well-chosen words."   —*Booklist*

"If you've not heard of Perry, he is a real find: Insightful, creative and yet down-to-earth, a writer with poetry both in his soul and in his writing. Read *Coop*."
—Barbara Rixstine, *Lincoln Journal Star*

"Sometimes funny, sometimes touching, sometimes both."
—Lisa McLendon, *Wichita Eagle*

"The charm of this writer grows with each book. Perry blossoms with *Coop*. . . . [His] gift is applying the small events of his being—and some not so small—to the larger truths of humanity."   — J. E. McReynolds, *Oklahoman*

HR 04 26 2023 0223

Coop

Also by MICHAEL PERRY

FICTION

The Jesus Cow

The Scavengers

NONFICTION

Visiting Tom: A Man, a Highway, and the Road to Roughneck Grace

Truck: A Love Story

Population: 485: Meeting Your Neighbors One Siren at a Time

Off Main Street: Barnstormers, Prophets & Gatemouth's Gator

AUDIO

Never Stand Behind a Sneezing Cow

I Got It from the Cows

The Clodhopper Monologues

MUSIC

Headwinded

Tiny Pilot

Bootlegged at the Big Top

MICHAEL PERRY

Coop

A Year of Poultry, Pigs, and Parenting

HARPER PERENNIAL

NEW YORK • LONDON • TORONTO • SYDNEY • NEW DELHI • AUCKLAND

HARPER ● PERENNIAL

A hardcover edition of this book was published in 2009 by HarperCollins Publishers.

P.S.™ is a trademark of HarperCollins Publishers.

COOP. Copyright © 2009 by Michael Perry. Photographs © J. Shimon and J. Lindemann. All rights reserved. Printed in the United States of America. No part of this book may be used or reproduced in any manner whatsoever without written permission except in the case of brief quotations embodied in critical articles and reviews. For information address HarperCollins Publishers, 195 Broadway, New York, NY 10007.

HarperCollins books may be purchased for educational, business, or sales promotional use. For information please e-mail the Special Markets Department at SPsales@harpercollins.com.

FIRST HARPER PERENNIAL EDITION PUBLISHED 2010.
REISSUED 2015.

Photographs © J. Shimon and J. Lindemann
Designed by Leah Carlson-Stanisic

The Library of Congress has catalogued the hardcover edition as follows:
Perry, Michael.
   Coop: a year of poultry, pigs, and parenting / Michael Perry.—1st ed.
   p. cm.
   ISBN 978-0-06-124043-0
   1. Perry, Michael 1964– 2. Perry, Michael, 1964-—Family. 3. Farmers—Wisconsin—Biography. 4. Farm life—Wisconsin. 5. Rural families—Wisconsin. 6. Wisconsin—Biography. I. Title.
   CT275.P5767A3   2009
   977.5'043092—dc22
   [B]
                                                                    2008043832

ISBN 978-0-06-124044-7 (pbk.)

23 24 25 26 27 LBC 17 16 15 14 13

For the tiny pilot

AUTHOR'S NOTE

Recently I had an apparently deep thought. I scribbled it down quickly so that it might not escape the loosely woven sieve that is my brain. I then spent untold hours polishing the scribble until it was an aphoristic gem of original profundity. Shortly after that, I revisited an essay I had written seven years previous only to find the exact same observation, typed nearly word for word. I have reached that point in my life where every other thing I say is something I've said before. For instance, I regularly catch myself trying to describe the scent of dirt. I am repetitively waylaid by my affection for particular words. (The modifier "little" pops up like Whac-A-Mole in my rough drafts and is never completely eradicated.) Further amplifying the echo, bits and pieces of this book were worked out in shorter, previously published pieces some readers may recognize. I like to imagine this reflects the development of a "certain literary style" when in fact it is more likely I am developing "certain nervous tics." To say nothing of engaging in a freelance writer's favorite sport: *recycling*.

When I'm not repeating myself, I'm contradicting myself. For instance, in the book *Truck: A Love Story* I stated that my father never allowed us to have toy guns; lately I recall we were allowed to keep a pair of realistic-looking squirt guns given to us by a relative. I once wrote of a cow called Angie only to find out her real name was Aggie. (Other mistakes of bovine nomenclature have likely been made—I tell you this because the cows cannot speak for themselves.) The "ten-day" Wisconsin deer hunting season is only nine days long, no matter what I wrote in my most recent hardcover. Sometimes readers point out these contradictions. If they are offered in collegial spirit (we are in this together) I am nearly always happy to post them on my Web site as evidence that my head and feet are a matching set of clay.

Finally, since I believe the term *nonfiction* depends above all on a reader's trust, I must disclose a few intentional partialities. I often change names to give friends and neighbors a veneer of privacy. I use the term *recently* with some latitude, and for the sake of forward motion I reference the lambing season of 2006 in the context of 2007. Finally, when I write of the church of my childhood, I do so knowing that some will object to the portrayal as either too critical or too benign, and all will find it incomplete, especially when the sect is small and history is spare.

I am grateful for anyone who reads my writing, even—or especially—with a critical eye, and one phrase never suffers from repetition: Thank you, reader.

Coop

PROLOGUE

At the earliest edges of my memory, my father is plowing, and I am running behind him. I see my feet, going pat-pat-
pat over the soil, I see my father, left hand on the wheel, right
forearm braced against the fender, head turning back to check
the depth of the plow, then forward to gauge his progress. The
soil is red and sandy in the high spots and dark and loamy in the
low spots, where it curls from the plowshares like strips of lico-
rice, leaving me this square, shin-deep trough in which to travel.
I trail the sound of the little tractor, so close to ground I can hear
the soft plop of the overturned clods. Now and then the plow
slices the soil so cleanly that a chubby white grub drops into the
furrow, unscathed. The grubs are translucent white, their black
guts dimly visible, as if through rice paper. Grackles and cow-
birds flock the plow, pecking through the new-turned dirt. The
grub will not last long. There is my father on his underpowered
Ford Ferguson, and there is me trotting right behind him, and
there is God above, looking down as I run the straight groove of
the furrow, my life laid out on a line drawn in the earth.

In the company of our six-year-old daughter Amy, my wife Anneliese and I have recently moved to a farm. I would like to present some sort of grand agrarian charter, but the whole deal is predicated mainly on the idea of having chickens. We are not alone in this: These Troubled Times seem to have precipitated a fowl renaissance. Mail carriers labor under a groaning load of multicolored hatchery catalogs, the latest issue of *Backyard Poultry*, and perforated containers that peep. Drop the term *chicken tractor* in mixed company and behold the knowing nods. The online world is alive with Subaru-driving National Public Radio supporters trading tips on eco-friendly coop construction and the pros and cons of laying mash; my NASCAR-loving brother-in-law tenderly minds a box of chicks beneath a heat lamp in his garage; my biker bar bouncer–turned–Zen Buddhist pal Billy and his wife the certified nursing assistant are building their second backyard coop with an eye toward expanding into "ornamentals." Anecdotal evidence to be sure, and a drop in the Colonel's bucket, but something is afoot. The subject of chickens was raised between my wife and me fairly early in our courtship, and has sustained us. We are enthused by the idea of fresh eggs, homegrown coq au vin, and (at least until butchering day) a twenty-four-hour turnaround on the compost. In addition, it is my long-standing opinion that entertainment-wise, chickens beat TV.

Our move is also family-driven. We are assuming responsibility for a farmstead previously owned by my wife's mother. Faced with an unexpected relocation, my mother-in-law wants to keep the place in the family. And in a bit of a flip, we are moving from a northern village to the country in order that my wife might be closer to the university where she sometimes teaches. This will save gas and time, although that glow on the horizon is a mall, and whenever we notice a Prime Commercial sign forty acres nearer, we review our escape plan from a place on which we have yet to pay the property tax.

We are also going rural in the hope that we might become more self-sufficient in terms of firewood, an expanded garden, and perhaps a pair of pigs. Whether through prescience or too much nervous reading, we have developed a low-key doomsday mind-set regarding the imminent future, and believe the time has come to store up some potatoes and teach the young'uns how to forage. Amy can already identify a coyote track, and I intend to see to it that she carry the phrase "slop the hogs" a generation further. To an extent, my wife and I are acting on positive recollections of our own childhood—I was a farm kid from the age of two, and much of my wife's childhood was spent on a farm just one valley over from the spot to which we have moved. But I hope we don't burden Amy with the idea that living outside the city limits is an inherently pious act. That *rural* equals *righteous*. As a country kid, I took a while to round the bend on that one, but thanks to a blend of peak oil posts, Kwame Anthony Appiah's *Cosmopolitanism*, and a week spent buying groceries at a bodega in Bushwick, I am well on my way to reconstructing all residual prejudice. Let's hear it for sensible urbanism. Whenever I catch myself waxing unctuous on the subject of getting "back to the land," I think of the Parisian Jim Haynes, who has said city dwellers protect the ecological balance of the countryside by staying away from it. And then there is Ben Logan and his touchstone book *The Land Remembers*, in which he wrote, "All around were reminders that the land was more important than we were. The land could do without us."

This new place of ours is not far from where I was raised—you can throw in your *Essential Steve Earle*, go the speed limit, and arrive well before "Continental Trailways Blues"—but it feels far. I arrived at our New Auburn farm in diapers and departed at the age of emancipation. Spent a dozen years away, then returned to live in the village itself. Twelve years into my second citizenship, I was settled and content within the rough bounds of an area framed largely in terms of memories and gratitude. I was in the fire department, I never missed Jamboree Days, I shot the breeze with folks in the post office lobby, and I caught up on politics at the village dump. Over half my life, I have happily punctuated my return address with the same zip code, the nicely rhythmic *five-four-seven, five-seven*. New Auburn, Wisconsin. The idea of forwarding the mail leaves me queasy and blue.

It would be sweet to noodle along in this minor key, but I'm stopping now because having pried my eyes from the compass mounted in my navel, I see the world is gray with the dust of diaspora and displacement. Any given moment, put your ear to the earth, and you will hear ten million shifting feet. Vicious herdings and abject decampments, perpetually under way and commenced at the business end of boots, bullets, and bulldozers. Whereas we have simply eased down the road a piece.

Our new farm is on hilly terrain, and when I come up the driveway I find myself reflexively checking the windshield to verify that my state park stickers are up-to-date. Having been raised a swamp and flatlands boy, I view all topographical rumples as exotic. I am more attuned to brush than vistas. When I walk the ridge, I can't help but remember all those Louis L'Amour books I read as a kid, where some fool skylines himself and is culled by a Sharps 50. When we were tots, my friend Harley was orphaned by a tractor when it pitched on a hill and crushed his father, and I have conflated hills and farms with danger ever since.

The ridge runs from the house at an angle straying off the east-west axis. As a result, the first time I came here I got my directions wrong, and I'm still trying to rewire. My disorientation is exacerbated by the fact that some of the outbuildings align with the ridge, while others are set square to the four directions. I am regularly startled by the apparent repositioning of sunrise.

I am nervous about some of the newer houses nearby. They tend to be grand. Again, I must disabuse myself of reverse classism—orthopedic surgeons want chickens too—but I cherish a regular salting of trash heaps and trailers, signaling as they do that the neighborhood will tolerate bad luck and alternative preference. It soothes me that our screen door does not latch, and you can entertain yourself during downtime by dropping a glass marble beside the toilet and quietly betting as to whether or not it will clear the tub surround and roll clear out into the kitchen.

I first perceived my father as a farmer the night he drove home with a giant lactating Holstein tethered to the bumper of his Ford Falcon. There was no cart, just a cow on a rope. And Dad, motoring real slow.

It wasn't the sort of scene you'd see on the cover of *Hoard's Dairyman*.

We went to fetch the cow after supper, from a farm some three miles distant. Owning neither truck nor trailer suitable for transporting the beast, Dad chose the Falcon—a station wagon model with a nifty roll-up rear window and a naughtily noisy Hollywood muffler. This being a momentous event, we kids—six of us at the time—clamored to come along, and we made a carful. That said, I have not discovered any photographic documentation of the event. Perhaps a couple of the younger children remained at home with Mom. When we arrived at the farm, a man led the cow from the barn on a rope halter. Dad cinched a granny knot around the bumper and set off at a stately pace. My smaller siblings knelt facing rearward in the seats, while my brother John and I sat backward on the lowered tailgate, legs swinging on either side of the Holstein's raccoon-sized head.

Three miles is not so far as the crow flies, but it is a fair distance when you are towing a Holstein behind a Ford Falcon full of jabbering rug rats. Dad rode the clutch and kept one eye on the mirror. The cow stubbed along reluctantly at first, all straight-necked and flat-eared, but eventually she calmed and found her road gear. For the balance of the journey she shambled along easy, following her nose through a faint blue haze. That Falcon burned a little oil.

As we neared home that night, John and I bailed off the tailgate to run alongside. The other kids hung out the windows, hair in the wind, arms dangling, all eyes on the cow. We tripped along through the lowering salt marsh mist, that sweet-brine scent still imprinted on my consciousness to a depth that zips me back to the far western corner of Sampson Township at first whiff. How happily and goofily we skipped, the scratch and scuff of the Holstein's hooves echoing among the roadside pines. Mom was in the house at the sink when we broke into the open, the blue station wagon rolling turtle-slow and trolling a lumbering milk cow through the lowering light and fireflies rising, John and I cavorting like redneck leprechauns, our father up there at the wheel thinking who knows what but proving that even the humblest man is capable of spectacle.

Lately—when we are driving to piano lessons, or washing dishes, or stacking firewood—Amy has taken to saying, "Tell me a story from your child-hood." She consistently uses the same formal phrasing (clearly enunciating "child" and "hood"), and I am tickled and troubled by the idea that to her mind my "child-hood" has achieved epoch status. Of course I remember asking my parents the same question, and I remember my fascination at their stories of boxing matches on the radio, ice delivered to the house in sawdusted blocks, and ladies who cooked with lard. To a child whose first retrievable memories are pegged just prior to the Summer of Love, it seemed remarkable that my parents were born into a black-and-white newsreel world, with still shots provided by my mother's Brownie box camera, still stored in a long narrow closet upstairs at the farm.

"Tell me a story from your childhood," says my daughter, and because I remember nothing before the farm, I tell her the first image I see when I think of spring is yellow dandelions in green grass beside a red barn. Ours was a classic Wisconsin homestead: white house, the red barn, a handful of simple outbuildings. All constructed from jack pine by second-generation Norwegian immigrants in the wake of the lumbering boom. The farm stood at the southern edge of what historians refer to as the cutover region. At one time the cutover was thick with century-old white pines, but by the 1900s they were long gone down the nearest river. As a short-term fix for the suddenly finite lumber supply, the loggers further stripped the land of lower-grade hemlock, cedar, and hardwoods. After the lumberjacks departed, the government encouraged farmers to settle the area, but the denuded sandy soil of the cutover was poorly suited to farming. Although there were many successful operations scattered throughout northwestern Wisconsin during my childhood, I came of age thinking of farming as a tough gig in which folks scrabbled and hung on—I was surprised in later life when I spent time in the Coulee Region to the southwest and discovered that farm families numbered among the prominent and well-to-do.

"Tell me a story from your childhood," she says, and I tell her of waking on summer mornings so socked with fog I could make myself believe the world had fallen away to leave our isolated farmstead floating through space—and how that illusion was gently undone by disembodied voices drifting in through the mist as our neighbors to the north called their cows to milking: "*M'bawssss . . . M'bawssss . . .*" I tell her how those same neighbors—a pair of bachelor brothers named Art and Clarence—still took the Norwegian newspaper and spoke of my brothers Jack and Jed and Jud as "Yack and Yed and Yud." I tell her of old men with knuckle stubs where fingers should be, and if you inquired after the digits the story invariably involved firewood and a homemade buzz saw. I tell her of Mr. Kenner up the road who can still tell you about the Indians who blocked the narrow bridge spanning Beaver Creek and charged a cut of tobacco for passage. I tell her of winters when we kids wobbled through the fields on our skates to play pom-pom-pullaway on a frozen drainage ditch carved through the middle of the woods by men serving in the Civilian Conservation Corps during the Great Depression. I tell her how we learned to milk cows and how we did flips from the haymow rafters, whirling and dropping through the cold winter air into the giant pile of chopped cornstalks below. Once I overrotated and slammed my knee into my nose. I remember the warm blood flowing out both nostrils toward my ears as I lay on my back. My brother John looked down at me without alarm. Squeezing my nose and looking up past him, I noticed that the nails protruding through the roof were furry with frost.

"Tell me a story from your childhood," she says, and I simply do not know where to begin.

My father recently joined the community choir. Sounds innocuous enough—sweet, even—but my immediate reaction was to phone my brother John and ask if he thought Dad might be smoking reefer. Four decades I've known my father, and he has led an avowedly quiet life. He works hard, he works quiet, he works above all to avoid any public act more conspicuous than renewing his driver's license. And now suddenly he's out there on tour (Chetek . . . Bloomer . . . Sand Creek . . . it's all a crazy blur), ascending the risers to raise his voice in public. For the *Christmas* concert, no less. My father is deeply devout, so on the face of it singing about the birth of Christ seems a natural match, but ours was a church so fundamental that the December holiday was banned as dangerously contrived pagan silliness— although my parents would certainly state their case in terms more demure.

So when Mom told me Dad had up and joined the choir, it felt like a flip-flop on one of those *How To Tell if Your Child is Using Drugs* moments. You've seen the bulleted lists: *change in usual activities; change in friends; new hobbies; drastic change in personality; change in clothing choices* (forty years in overalls, and there he is doing *fa-la-las* in a white button-down and green bow tie).

Seriously. I'm gonna check his sock drawer.

When my father and the rest of the traveling choir perform at St. Jude's, I attend with Amy in tow. St. Jude's was the Catholic church in New Auburn. I say *was* because a regional bishop shut it down en route to earning a significant promotion. That was hard to take, even for a lapsed Protestant like me. When my brother's wife was killed seven weeks into their marriage, St. Jude's opened their doors for the funeral without regard to affiliation. And for years now, whenever a Lutheran or Methodist event overflows the capacity of its respective church, the whole production shifts to St. Jude's.

In closing the door, the bishop cited economic concerns. It's always tricky when Men of God wield calculators. He showed up to do the job himself, I'll give him that. He looked pinched but resolute. I attended the final service out of respect for my neighbors, as it was their blue-collar tithing that paid for the shingles, the pews, and the bishop's remarkable matching cap and bathrobe. His decree has been ameliorated somewhat by the fact that the St. Jude's auxiliary continues to make the church available for weddings, funerals, and public events—including the community choir Christmas concert.

Huffing and stomping into the foyer from the frozen parking lot, we pack into the pews, cozy in our coats. We all do some swiveling, looking around to see who's here and who's where. On the stroke of the hour, the choir files out, the women in their dark green robes, the men in their white shirts and green bow ties. You see a lot of gray heads up there, and some cautious climbing of the risers. The men always look scrubbed up pretty good, with fresh comb tracks in their hair. Somewhere someone is still selling Brylcreem. The women look pleased and pleasant. My mom—wearing her bifocals and with her white hair up in a bun—is there, and so is my sister-in-law Barbara. Barbara's husband—my brother John—is in the back row stage right, his face windburned and bearded. Dad is also in the back row, stage left. Most everyone has a trace of kindergarten recital stiffness in his posture. The conductor is an animated fellow given to preacherhopping during the rhythmic parts, and given to hand-waving during the hortative parts, but you grant him latitude because he has himself quite a job there, summoning hosannas from a pack of mannered Midwesterners. During warm-ups I note he sports a scandalous earring, but then he *is* the leader of the band.

And so they sing for us. I make it to the Christmas concert about every other year, and am always enchanted with the homemade joy of it all. I think of these neighbors getting their work done and hustling through supper so they can make rehearsal on time, giving up their evenings in, their television shows, their early-to-bed. Doing it as fall becomes winter, fighting the first snowy roads. Memorizing their lyrics and learning their parts, with no expectation of remuneration beyond smiling faces and afterward, coffee and cookies. I see the dump truck driver raising his voice to the ceiling panels, hear the administrative assistant weaving her harmony with that of the farm wife, and think at that very moment of the googolplex infinitude of electrified screens and bangety-bang speakers blasting away at the world, and we are blessed indeed to be in this small space, with our neighbors singing for us. The music swells from worshipful chorale to hand-clapping swing, up and down, the tempo of the piece following the tempo of the tale. The music builds and builds, soaring to the point where I am rocking in my seat and bobbing my head some when my gaze shifts to Dad in the backmost row, and it hits me as a soft shock that he looks small up there. Still the bright eyes, still the wrestler's physical alacrity, but he is favoring one knee, and after years of little change, his hair has gone gray and gone thin. Right around his pop-out ears—the ones Mom swears she never noticed until someone mentioned it—the hair is tufted and a tad askew, so that when he peers through big glasses to see the music and sings with his eyebrows raised like he's been surprised by the next note, he's all absentminded-professor-looking. But more than that, the oversize glasses and the green bow tie render him childlike, even as I realize: that's my dad, becoming an old man.

Here on the new place there is much to do. Most of our things are still in boxes. A goodly bunch of the rest of our things are still in the New Auburn house. As far as self-sufficiency, Anneliese and I are doing our best to aim low—some eggs, some pork chops, some firewood; if we can raise just a portion of the heat and groceries by our own hand this first year, it will be an improvement on years previous. There is also the small matter of me nattering on about how I intend to be a subsistent man of the land even as I spend a quarter of the year ramming around the country with a trunkload of books. This is hardly proper behavior for a father, a husband, or a husbandman.

Also, we have a baby on the way. Due in early spring. Anneliese has proposed the tot be birthed right here in the crooked old house. The first time she brought this up, I laughed in a *Hahaha! Good one!* sort of way, and then, seeing her gaze harden ever so slightly, I said, well, yes, sure we could discuss that, sure! *Hahaha!* This time the *Hahaha* was a little higher pitched.

CHAPTER 1

This morning while splitting wood, I attempted to clear my left nostril using a rustic maneuver known as the "farmer snort" and misfired badly. My Eustachian tubes have yet to assume their former diameter. I bubble-gummed an eardrum, shot fizz out both tear ducts, and may have permanently everted one eyeball. I believe I sprained my uvula.

The act of blowing one's nose without benefit of Kleenex is a skill appreciated along a wide spectrum of background and endeavor—synonymous terms include *fisherman's tissue* and *air hanky*. But I grew up calling it a farmer snort, because them's my people, and them's who taught me.

The lessons were not formal. Watch and learn, learn by doing. Same way I learned to spit. Although Dad nearly derailed me there. We were walking from Oliver Baalrud's barn to the house for a lemonade break between unloading hay wagons when I puckered up and gobbed a stringer down my miniature bib overalls. I'd been watching Oliver all day. He was a diminutive Norwegian and an accomplished spitter—not big tobacco-ey streamers, just frothy little pips, but he did it constantly, and the flecks flew sharp and straight. Gosh, it was just the neatest thing. He could do it while talking, pitchforking, or backing up the tractor. I wasn't even in kindergarten yet, but I was trying to march beside my dad like a little hay-making man, and I guess I figured spitting would be the thing. When I goobered on my bibs, Dad didn't break stride. Just looked down and said, "Don't spit until you know how."

Boy, that set up a conundrum. How you gonna *learn* if you don't *do*?

I guess I got around it. Later, when I lost my milk teeth, the new set came in with a pretty good gap. This helps with the aiming, and has a rifling effect. Given a gift, you work with it. I can sit in the kitchen and knock a horsefly off your doorknob.

Executed smartly, the farmer snort is a source of transcendent clarification. In short, it really lightens your head, and consequently your day. Conversely, a snort misplayed can put a serious crimp in your karma. As with most things in life, your odds of success improve through focus and rehearsal. Determine your dominant nostril; visualize success; think *through* the snort— that sort of thing. I have encountered people who claim to be able to perform a hands-free double-barrel farmer snort. I am skeptical and not about to hang around while they prove it. The first time I saw my wife farmer-snort, I felt a renewed flush of affection and thought, *Now there is a woman who can endure.*

I split wood because I boldly predict that next winter will be cold. Climatic creep notwithstanding, one retains the long johns.

Ever since I departed the home farm some twenty-four years ago, I have warded off winter with a twist of the thermostat. Now that we've moved to the farm, we have backup electric heat, but the preponderance of warmth in our house is generated from within a square steel box in the living room, and the thing must be fed. I have this spot where I like to chop; it overlooks a swale that breaks into the valley below. Now and then I pause to unbend my back and absorb the view. I feel hardy standing here, muscular and flush with the full pulse of labor. There is sweat in my stocking cap, my typist palms are nicely sore from gripping the ax, and the splintered wood at my feet is tangible evidence of my attempt to provide for our little family of three. Our farmhouse—just up the slope from where I stand—is a mishmash of remodels and add-ons clad in scuffed aluminum. The windows are uncoordinated, and the floorboards are straight out of a carnival funhouse. Moving from the kitchen to the living room, you step up a four-inch riser; keep moving on the same plane around the central wall, and you will circle right back to the riser, having never stepped down. This creates an M. C. Escher effect and helps explain why my mother-in-law refers to the bathroom hall as "that wheelchair ramp." But at the heart of the structure is a log cabin built in the 1880s, and it is dead solid. A few of the hand-squared logs are still visible along one side of the living room. They are the width of a boar's back. Pausing ax in hand to gaze off across the territory, I picture myself as some austere pioneering backwoodsman on the order of Abe Lincoln—albeit dumber, stubbier, and unlikely to alter the course of human events, unless you count snoozing at a stoplight.

It's a good day for splitting. Fifteen below at dawn, and even in mid-afternoon the oak is frozen tight. Wet wood split in summer absorbs the ax with a punky *tunk!* You spend half your time wrenching the sunk blade free; the beveled steel cheeks press out a watery froth. Today at subzero, nearly every stroke terminates in a crisp *ker-rack!* The halves part neatly, releasing a scent like musk and cheese. The exposed wood is laced with crystals of ice that refract the sun and salt the grain with an interstitial twinkle.

I split a while, then stack a while.

The baby is due in early April, just over three months from now. After a slow start, I am astounded at the speed with which Anneliese's belly is growing. Whatever is in there, it is a kicky little creature, and prone to nocturnal hiccups. Nearly every night when we lie in bed, Anneliese's midsection begins to lurch sharply and at measured intervals. It is my understanding that the sensation is equivalent to the baby playing foosball with Anneliese's innards. It is tough to drift off with a miniature single-stroke engine boing-boinging between your liver and bladder, and compounding the problem, ever since Anneliese became pregnant, she has been struggling with insomnia.

We do not know if we are having a boy or a girl because we have had no ultrasound, and barring some pressing sign or symptom, will not have one. Anneliese is defiantly self-directed in these matters, relying on a coterie of friends covering the spectrum from pagan shamans to a home-educated evangelical Christian nutritionist. I have gone to the medicine cabinet seeking an aspirin and come up with powdered kelp.

I am nervous about the baby business, but this is hardly news.

I am nervous about everything—from last year's tax calculations to the blinking light on the answering machine. Discussions in the arena of health care do not always go easy between Anneliese and me. Thanks to the holistic curriculum provided to me by the University of Wisconsin–Eau Claire School of Nursing during the years I thought I knew what I wanted to do for a living, I am willing to give consideration to a wide range of alternative treatments, but remain decidedly partial to Western medicine. In part, this comes from my unwillingness to swim against popular tides, coupled with extreme trepidation about doing the wrong thing. My wife does not labor under the same unattractive state of wimpitude. In a marriage that has so far been everything I hoped it might be, our most difficult—even heated—discussions have been about medicine. Having said that, regarding neonatal issues, she has several advantages, chief among them being that (A) she is carrying the baby, and (B) since we are paying for prenatal care and delivery out of pocket, I am happy to go along with the economically attractive elements of her program.

Economy is as economy does, and having observed my progress at the woodpile for a month now, Anneliese has lately begun lobbying for a wood-splitting bee, in which we invite the neighbors and get the whole works done at once. She says I shouldn't be toiling out there all alone. I suspect she has also calculated my ax stroke to BTU ratio and fears that by next February we'll be busting up the last of the kitchen chairs for kindling. "We can make chili," she says. "Terry can bring his splitter, and we'll get it all done in one day." It's a good idea, and indeed, our neighbor Terry has offered to share his gasoline-powered wood splitter. It's a smooth little machine with a small engine and hydraulic ram mounted on a steel I beam that rides on a set of trailer wheels. You just hook it behind your tractor or pickup and tow it to wherever it's needed. Once the wood chunks are placed on the I beam, the operator moves a lever forward and the hydraulic ram pushes the bolt of wood against a piece of beveled steel the shape of an ax blade. The hydraulic pressure is slow but inexorable, and even the toughest knot-bound snag comes apart into manageable pieces. The first time I saw an automatic splitter I was a kid, and it was an overbuilt homemade contraption. Nowadays you can pick 'em up at the Farm & Fleet, all painted and shiny.

The alternative is to flail away madly with a splitting maul, sweating like an overheated stevedore and likely working up a stellar case of carpal tunnel. My wife is right. My pecking away solo is silly, and my left arm has been numb for a month in a dermatome representing the ulnar nerve. But I'm not out here to be efficient. I'm out here to clear my head. To feel the ax rise and fall, to blow the breath out on the downstroke and drive through the bolt. To bring the blade down dead center and see the wood explode—the satisfaction is, I suppose, similar to what a golfer feels when connecting with the sweet spot. When you hit the wood just right, it doesn't feel like you've dealt a blow, but rather that the steel head has floated through the wood. When you miss, on the other hand—when you deliver a stroke that ends in a dead wedge, or kicks out sideways and caroms into the dirt—you're often left feeling disoriented, sometimes with the same numb-palm sensation as when you swing early and hit a baseball off the tip of the bat.

I want to split wood by hand for the same reason I want to have pigs and chickens. You want to eat meat, you raise an animal and kill it, or at the very least steal its eggs. You want to stay warm, you knock the wood into little chunks. Beyond that, there is the idea that primitive, meaningful work delights the mind. When undertaken in the absence of compulsion, I should say. One regularly finds that praises of an honest day's work are frequently sung by people in clean clothes. I am thinking more along the lines of the character created by Jim Harrison in the novel *Returning to Earth*, who says his blue-collar jobs "kept me grounded in actual life." I am being selfish about the wood. Proprietary, even. When I chonk a chunk into the firebox, I want to stand back and claim it a bit. A man, being a man. Providing.

You take that ax in hand, and it frees your mind. Of course, too much dreaming and it will also free your toes. I am regularly dramatic with my wife about accumulated pending deadlines and backlogs and time spent on the road, only to have her look out the window and see me there chopping when I should be typing. In proposing the firewood bee, she is being eminently sensible, and that is where we part company.

If she brings it up again, I shall tell her I am freeing my mind.

The centerpiece of my parent's farmhouse is a stout Monarch Model 3755D wood-burning range. Carl Carlson, the man who homesteaded our farm, bought it as a wedding gift for his wife Charlotte when they were married in 1920. When he vetted the stove prior to purchase at the Farmers Store in New Auburn, the salesman demonstrated its durability by jumping up and down on the open oven door. Product testing of equivalent rigor is unimaginable in our stamped-tin age and will furthermore get you Tasered at the Best Buy. Behold the mighty nation gone slack.

When the Carlsons installed an electric range, the Monarch was banished to the basement. Mr. Carlson reattached the stovepipe to the chimney, and Mrs. Carlson used the stove to burn garbage and make lye soap. By the time my parents arrived, it was coated with rust and dinge. Looking as ever for ways to pinch pennies, Mom and Dad decided to move the stove back upstairs and use it to supplement the furnace heat. In order to do so, Dad had to disassemble it—unbolting the warming ovens, detaching the water jackets, and generally breaking it down into the smallest components possible. It was still a prodigious project; I have seen the original catalog materials for the 3755D, and it weighs 442 pounds—that's a lotta microwaves.

A neighbor came to help with the lifting, and once the stove was reassembled upstairs, its squat bulk anchored the entire first floor. Mom cleaned it up and rubbed it down with blacking, and although the shiny bits were dimmed and pitted, they did take a polish, and the blue *Monarch* logo still scrolled beautifully across the white porcelain enameling of the oven door. She rarely baked in the stove, but we often came in from wood-gathering expeditions to the scent of smoked ham and vegetables in a cast iron pan that had percolated on the stovetop all day long, and as we ate, our caps and mittens dried in the warming ovens flanking the central stovepipe and its butterfly damper, which reminded me loosely of the Batman logo. On cold school mornings, we tussled to see how many of us could plant our hindquarters on the warm oven door.

In winter the day began with the sounds of Dad building the fire. From our beds upstairs we'd hear the hinge-squeak of the firebox door and the clank of the lids and center plate as he lifted and set them aside. Next came the rolling rumble of the grates as he shook them free of ash. The grates were rotated by means of a detachable cast iron handle fitted over a stub of square shaft protruding from just above the draft door, and were concave on one side and convex on the other; each time Dad twisted the shaft, a nickel-plated indicator countersunk on the front of the stove slid back and forth, alternately reading WOOD or COAL. When the grates were clear and returned to the WOOD position, he detached the handle and stowed it in one of the warming ovens. Even this action had its own distinct sound: the tinny scrape of the handle sliding back in place and the *clunk* of the warming oven door stakes dropping into their pockets. If the ash pan was in need of emptying, we'd hear the gritty rasp of it being pulled from the square steel pouch where it nested beneath the grates. Then the front door would open and close and the house would go quiet while Dad walked to the garden and flung the ashes across the snow, where they left a skid mark like a miniature thundercloud run aground.

Then he was back inside, and even now I can summon the image of him downstairs alone, the day's work in mind, the simple ceremony unfolding. The crumple of the newspaper as he packs it in above the grates. The careful placing of the kindling, and then a few larger sticks of wood to catch and grow the first flames. The lids nesting flush with the stovetop when he replaces them, fitting their receptacles with jigsaw-puzzle precision. The scratch of the match against the sooty interior of the firebox door,

Dad ducking his head to light the tinder, the prayerful stance of it, him on one knee and blowing gently at the flame in the predawn darkness, and us his family still abed. Mom and Dad still use the Monarch. It sits right where it has since the day it came back upstairs, just feet from the dining room table. Even today, when we kids gather as adults, someone (or sometimes two of us, if personal dimensions allow) winds up perched on the woodstove door. We sit there even when the weather is warm and there is no fire. Something more than warmth draws us to the stove, something having to do with memory and the center of gravity.

When it comes to parenting tools, it's tough to beat a woodstove. Pick up your room, we say, because . . . because . . . *never mind what Daddy's room looks like! Daddy is not the subject here! Daddy is a full-on poster boy of undiagnosed behavioral disorders!* Be nice to everyone, we say, because . . . because . . . *Yes, even that lady who "waved" at Daddy in the Wal-Mart parking lot . . . and the snotty little ingrate who stole your beach bucket . . . Why? Because . . . because . . . well, because passive-aggressive is the only way to roll, sweetheart.* In other words, how does one convey cause and effect to a six-year-old?

By having her haul firewood, that's how. You wanna lie around toasting your tootsies, darling daughter? Then get out there and lug some cellulose.

In a sense, my siblings and I lucked out. Dad logged every winter, which meant the sawmill came most summers, leaving behind a giant pile of slab wood, which didn't need to be split—just sectioned up and stacked. We called these slabwood chunks *schniblings*—a word we learned from a neighbor up the road. I'm not really sure how you spell *schniblings*. Most of the time we shortened it to *schnibs*. Even now as I type this, I fear *schniblings* will turn out to have been some outrageous ethnic epithet. If so, forgive me. I Googled it and Babel Fish'd it and came up with nada.

The downside of our method was that most of the trees Dad harvested were white pine, which has a burn rate roughly equivalent to Kleenex, so it took a mountain of slabs to keep the house warm. When it was time to "make wood," as the common phrase had it, Dad rounded up the troops and took us to whichever corner of the farm the sawmill had been set up in last. Sometimes we all climbed in the back of the pickup; sometimes we rode on a hay wagon to which Dad had attached side racks. A certain glumness prevailed when we were in the hay wagon because it was much larger than the pickup, and we were anticipating a marathon. While Dad was gassing up his saw, we kids began stacking slabs in a pair of sawbucks that cradled the wood in a bundle. With one side barky and one side rough-sawn, one end fat and one end knife-skinny (or thin in the middle so they snapped in two mid-lift), the slabs were splintery, unbalanced, and a hassle to handle. Yanking them from the every-which-way pile was like playing full-contact jackstraws. At Christmas, when we went to the city and stood on the carpet of Grandpa's split-level ranch and watched him fill the fireplace with uniform cylinders of papery-smooth white birch, I remember feeling what can only be described as firewood envy.

Back and forth we went between the sawbucks, alternately filling and emptying them as Dad ran the saw nonstop. We slung the chunks into the wagon or truck bed, stopping now and then to peer hopefully over the side racks. It seemed ages before Dad killed the saw, helped throw the last batch aboard and headed for the house. But the work had barely begun—the wood had yet to be unloaded and stacked in the basement. In later years Dad built a wood chute, but we used to just pull open a window and fling the wood through the opening. When we finished, the sill was battered and busted, and the window had to be held in place with a bent nail. Finally, we stacked the wood, often by increments after school. By the time the first snow fell, the basement was a warren of wooded corridors leading to the root cellar, chest freezer, and sump pump.

The penultimate step in the slab wood journey was the wood box—a large antique crate positioned directly adjacent to the solid Monarch. Once the wood had been stacked in the basement it came back upstairs one armload at a time over winter. By the time you made it upstairs, your biceps were aching and it was a relief to hear the noisy tumble of firewood spilling into the crate. It took a lot of trips to fill the wood box, but the following morning when we raced each other for the stove door, we had in some measure earned the warmth on our hindquarters.

Now that we have moved to the farm, poor Amy has come to understand this dynamic all too clearly. One of her daily winter-time tasks includes making the long trudge to the old granary across the yard where the dry wood is stored. Watching her load up her purple plastic sled and drag it slowly back to the house, I smile, remembering all the times Dad pried me from behind a Louis L'Amour cowboy book to do the same. More often than not, she goes willingly, if not gladly. If she sulks or fusses, I launch into an eye-glazing sermon, reminding her of how many times I have found her curled up in front of the stove with Dora the Explorer, and do you *know* where that warmth comes from, and let me tell you when *I* was little we had to go all the way out on the *back forty* to get *wagonloads* of wood, and, well, on and on it goes until Anneliese gives me the look normally delivered from the front pew by the wives of long-winded preachers, at which point I stalk off in a cloud of my own oration. Meanwhile, Anneliese explains to Amy that she is not just doing *her* chores, she is *helping the family*. Anneliese presents these lessons in terms any six-year-old can grasp, and sees no need to revise upward when—as is regularly required—she adjusts my own focus.

I recently stepped into the upstairs hallway just as Amy emerged from the bathroom cinched underarms to knees in a towel. As I watched, she dropped her head forward, wrapped her dangling wet hair in a second towel, twisted it turban-tight, and then—in a single unbroken motion—rose upright and flipped the tail of the towel back over one shoulder before scampering to her room. I stood stock-still, having just witnessed the future rocketing beyond my grasp. Of course I saw her mother in the movement, but I also detected a more universal womanliness, a posture of assurance. Were the moment to be rendered in neon, you would have this bright buzzing sign flashing *See Ya Later, Old Man*.

Amy is my *given* daughter. The term is not mine. A poet friend blessed me with it when I was trying to work my way around the word *stepdaughter*—a term I find serviceable by way of explaining the situation but utterly short of the mark when it comes to expressing the heart. Amy's father Dan lives in Colorado, and I am grateful to say that we get on well. As a matter of fact, we have just returned from a visit with him, his wife Marie, and their two toddling sons. Amy relishes the chance to play big sister, and quite rightly calls the boys her brothers without qualification or prefix. As for the adults, we are nearly four years into a relationship that is in some respects highly unusual, but ultimately exactly as it should be. We are sometimes complimented on how we have managed to skirt the mire, but not a one of us takes the situation for granted, and if the subject is raised, each will point to the critical contribution of the other three. In these situations only one person is required to bring the whole deal down, so: *Yay, team.* I am reminded of a party trick I learned as a child in which you set four water glasses butt-up in a square pattern and then weave the blades of four butter knives in such a way that the handle of each knife rests on an upended glass and the blades form a self-supporting grid in the middle. Once the blades interlock they will support a fifth glass filled with water. Remove any one of the knives and the whole works collapses, dumping the water. Amy is the water in the glass. Amy's father says we are "an anomaly relying on grace and friendship." Unfortunately, he is eloquent and able to express himself without resorting to party tricks as allegory. Additionally, he stands six-foot-seven, has all his hair, and can make raspberry coulis from scratch.

If there was any lingering alpha male tinder smoldering between Dan and me, I trust it was snuffed on the third night of this most recent visit. Thinking I heard a call from Amy's bedroom, I ran upstairs to check on her and found that she had taken ill. Based on her sad puppy eyes and chalky countenance, I determined she was shortly due to hurl. Grabbing a towel from the doorknob, I scooped her up just as the first blast blew. I caught most of it in the towel and hustled to the bathroom, where I wrapped a steadying arm around her and used my free hand to keep her hair from her face while the poor kid heaved in the toilet.

During the first false lull, she raised a plaintive cry: "What is *HAP-pen-ing?*"

And I realized: this was her first-ever upchuck session. Of course she'd spit up as an infant, but carried no memory of it. This was the first real deal.

"You're throwing up, baby," I said, projecting calm reassurance. "It's because you're sick. It's no fun. But it's OK. You'll feel better." She barfed again.

During the next lull, through tears she said, "This is a *really bad day!*"

When the vomiting stopped for good, I stood at the sink, running a cool rag over Amy's face. By now Dan had come to help. When I looked up into the mirror I saw him reflected behind me, dipping the towel up and down in the toilet while simultaneously flushing away the throw-up. "Little trick my mother taught me," he said when our eyes met, and I remember thinking, What are the odds of this moment?

Yay, team.

Amy is growing so fast. You think you hear that all the time, but I mean *growing*. She is six years old and the top of her head comes to the middle of my chest. No guarantees, but it would appear she is headed for the far side of six feet. Not only is Dan six-seven, he is the shortest of three brothers. His mother is an elegant woman of six feet. His sister—who looks just like Amy in her baby pictures—is six-two. Being so tall can be tough on a little one—even we sometimes grow impatient with her based on the age projected by her height as opposed to her actual chronological stature. And because we are homeschooling, we often forget how tall she is until she goes to dance class or swimming lessons and stands beside her peers. Still, we're taking the straight approach. We just tell her, Yes, it looks as if you will be tall—just like your beautiful grandma, and your lovely auntie. Sometimes in rehearsal for the future I lift Amy up and stand her on a kitchen chair. I am five-eight at my most optimistic, so I tip my head back, shake my finger at her from below, and say, "Go clean your room!" And then in unison we both say, "Just practicing!" and she always laughs.

Shortly after we return from Colorado, she gets chicken pox. Poor kid, this makes three times she's been sick since Christmas, when she broke out in red spots all over. It looked like measles. Being self-insured, we made the diagnosis using Google and my twenty-five-year-old nursing textbooks. I was up at 2:00 a.m., going back and forth between text and screen. Everything matched up, with the exception of some spots in her mouth. I dug a little deeper, and there in the text discovered scarlatina, which is potentially less hassle than measles. The buccal cavity lesions matched up, and so we called it.

The thing that caught me off guard was how helpless I felt when she was sick. Over the years my parents have taken care of many severely ill children, some of them terminal. So I've seen far worse than flu and scarlatina. But this would be the first time it was a child for whom I was responsible. Early indications are that I respond by going weak-kneed and turning everything over to Anneliese. In theory I support her in her determination to not be medically overaggressive, and I am even open to certain alternative therapies, but the minute the kid is symptomatic, I'm ready to run for the drops and pills. In this case, Anneliese held the line, and soon Amy was better. I tell you though, the next time I heard Greg Brown's song "Say a Little Prayer," it sat me right down.

We are still in the process of moving possessions from our New Auburn home, and with the real estate market in a funk, there is no sign of an imminent sale. I've already caused Amy to move thrice between the age of three and seven. When we met, she and Anneliese had finally settled into a home of their own after living in more temporary conditions in Colorado and Wisconsin. After Anneliese and I married, we sold their house, and they moved to mine. Now we're moving again. As a kid allowed to grow up in only one place, I wonder what effect all this moving will have on Amy. Sometimes she gets teary and says she will miss her New Auburn bunk bed. Sometimes she gets teary about what she calls her "Talmadge house" (she and Anneliese lived on Talmadge Street). So I am second-guessing our move when we take time from a hectic day to stop by the New Auburn house for another load of boxes, but as Amy waits for me to unlock the door, she looks up cheerily and says, "This is one of my hometowns!" She says it like more than one hometown is a good thing, and this lightens my heart.

Somewhere along my patrilineage—I think it may have begun with my great-grandfather Wyman—our family developed the habit of responding to childish or unrealistic demands with the phrase "One, two, three, *want!*" "I want a new *wagon!*" I might say, to which my father would respond, "One, two, three, *want!*" Translated, the phrase means "fat chance." Of course when Amy came into my life, I was eager to pass it forward. So now we are in Farm & Fleet, and as we pass the toy section she peels away from me. I turn to find her transfixed before a horrifying blister-packed horse-with-princess set. "Ohhh, it's *beautiful*," she says, turning her eyes up to me in that wide-open way that puts a lump straight in my throat. "I *really, really* want it!" She is heartbreaking in her sincerity. We adults work overtime to mask our desires, but a child just comes right out and tells you, and even when it's a horrific piece of plastic crapola—or perhaps *because* it's a horrific piece of plastic crapola—a child's willingness to so nakedly admit they *really, really* want it literally brings tears to my eyes. Unfortunately for Amy, the tears do not wash away my resolve. "One, two, three, *want!*" I say, and depart for the checkout line.

When I get there, I turn and realize she hasn't followed me. She is back there before the horse and princess, and she has her fists squeezed tight at her sides and her eyes shut and her little brow is furrowed with focus, and she is saying, over and over, "One, two, three, *want!* One, two, three, *want!*"

The poor kid thinks it's an incantation.

It is hard for me sometimes, to watch this girl—the one who loves to go rambling in my old pickup truck, the one who happily cuts up dead deer on the kitchen table, the one who lugs firewood—stare all googly-eyed at blond plastic princesses, but there you are. Sometimes all the academic feminism in the world can't compete with a chintzy tiara. And truth be told, her interest in the princess worries me less than her interest in the horse. The first thing Amy showed me the day I met her was her collection of plastic horses, and her fascination with everything equine has only grown since then.

I fear my daughter is Horse People.

You know Horse People. I am not talking about those loftily behatted julep-sippers of the Triple Crown circuit. Those are hors*ey* people. Whereas horse people are generally solid citizens with day jobs. You see them behind the counter at the bank, or working reception at the doctor's office, or hanging your drywall, and you don't suspect a thing. But their closets are full of Wranglers and pearl snap shirts, and their backyards are circumscribed with electrified white poly tape, and they will sometimes lapse into talk of snaffles and gymkhana, and somewhere out back is a round-nosed trailer with green windows. These are otherwise rational folks who nonetheless devote an unbalanced preponderance of resources to keep in their possession a large four-legged animal whose one big trick is the ability to transform overpriced hay bales into road apples. I understand I am treading in dangerous territory here, similar to offending cat people (let's skip right over ferret people, shall we?), but where I was raised, superfluous horses were known as "hayburners." I also admit I once had a bad experience on a horse named Warts.

I didn't give in there at Farm & Fleet. I may get the emotional sniffles, but I do not surrender. Having said that, I am aware that under the right circumstances a horse can serve as the intersection where joy and responsibility meet. In short, it is highly likely that I will one day own a dang horse.

In the meantime, she can have a guinea pig.

We have told Amy that the guinea pig will be her responsibility, and that how she executes her commitments specific to the pending *Cavia porcellus* will have a direct reflection on the possible future expansion of animal husbandry commitments up to and possibly including the acquisition of an equine division. We didn't put it in those exact words. Actually, I did, but Anneliese sent me off to sort socks before I really got rolling.

Amy doesn't know it, but we've already procured the pig. My sister-in-law Barbara is part of a multistate rodent rescue ring. I am not kidding, and if you are tempted to snicker or make jokes about the Underground Rodent Railroad, I should add that Barbara spends six months of the year driving a Mack truck and is licensed to argue cases before the United States Tax Court—the woman has a lot of ways to hurt you. I for one applaud her efforts on behalf of the order Rodentia, and intend to send a check to the appropriate foundation.

The animal in question came from Indianapolis. Aunt Barbara drove down there and picked it up herself. Again, I am not joking, and may I remind you, the IRS is not lumbered with the standard statutes of limitations. The guinea pig is currently residing with Barbara and my brother John until we can arrange the handoff. They live in a small log cabin up north, sharing quarters with a small herd of gerbils and hamsters, a large fish named Oscar, and our guinea pig, to be named later.

We are on our way to see the midwife. This will be my first visit. I missed the first one because I was on the road. My mother— who once worked as an obstetrics nurse—went in my place. We are driving north to a small town where the clinic is situated. There has been little snow. The countryside is mostly brown, and my mood is perfectly suited. I am quiet and grumbly. Things always look fine on the calendar, but then when the time comes, invariably I have six other things I'd rather do. Considering the circumstances, this renders my grumpiness pure selfishness. I have no excuse, only self-awareness.

The midwife rents space from a chiropractic center. We enter through a back door facing a parking lot and pass down a hallway past plastic skeletons and posters of the nervous system before we reach the small room where the midwife is set up. "Welcome!" she says. "I'm Leah." She greets me with a smile and a handshake, and hugs Anneliese. Leah looks athletic and earthy, and she smiles beautifully at Anneliese, oohing and remarking over her belly. When she shakes my hand her grip is strong and we both smile politely, but I see reservation in her eyes, and I know she sees the same in mine, because there are reservations in my heart. She knows from Anneliese that I am a skeptic. I am open to the idea of home birth because I love my wife and this is what she wants, but I am also bucky about the idea of delivering babies old-style if it is simply in service to some whole-grain earth mother sensibility picked up during a women's studies course in Colorado. As a former fundamentalist gone agnostic, I tend to dig in my heels at the first whiff of evangelism, whether it be deployed in the service of salvation, Girl Power, or the curative wonders of organic yams. There is also the frank issue of testosterone—four years in nursing school and three Indigo Girls albums notwithstanding, I am not purged of all chauvinism and not interested in achieving complete anemia. In short, a man likes to drive. Even when he's lost.

Leah asks if we would like tea. Amy chooses chamomile, and Anneliese chooses a prenatal brew combining herbs possibly capable of aligning the baby with the axis of the earth. I choose green tea, figuring its caffeine is the closest I can come to recalcitrance in this setting. Amy goes for a box of toys in the corner, and Anneliese and I sit beside each other on a small couch. On the low table before us I see a fanned set of pamphlets advertising a woman coming to town to speak on Bible prophecy, and right about then Leah produces a form and says she needs to update Anneliese's health "herstory." I can feel the voltage of my force-field cocoon ramping up.

While the kettle heats, Leah pulls her charts and papers and begins interviewing Anneliese. I am struck by how animated my wife has become. She pulls a list of written questions from her purse, and as she and Leah run through them one by one, I feel a twinge of something like jealousy wrapped in unease. She is gushing about the baby, about her body, about how the days and nights have been going. Even when she describes her problems with insomnia and lack of appetite, she seems palpably grateful for Leah's attention. It's bracing to see her come to life this way, to sit outside the circle of sisterhood and understand how this baby is inhabiting my wife beyond the womb. On the floor, Amy is playing with a reversible pregnancy doll. Flip the dress one way, the doll has a belly bump; flip it the other way, you have a mother holding a baby.

After checking and charting Anneliese's vital signs and testing her blood for iron (she is holding steady from the last visit but still a little low—this is one of the standards that must be met to have a home birth), Leah turns to me and asks if I have any concerns. I tell her I am in, but want assurance that if there is trouble, we pack up and head for the big hospital with the shiny lights. "Of course," says Leah, with no trace of defensiveness. "It is my responsibility to be honest with you if we are in a situation beyond my abilities." By the time she finishes reciting the list of risk factors that trigger "transfer of care," I am more settled.

We move to the examination booth. As Leah palpates Anneliese, Amy and I study a large wall poster showing fetuses at actual size along various stages of development. We find the six-month image—clearly recognizable as a human, but not much bigger than a Cornish hen. When we turn to look at Anneliese's belly and extrapolate, Leah is leaning over her wearing a stethoscope that projects from her forehead on a spring-loaded frame reminiscent of a phrenological measuring device. Using her forehead to press the bell against the skin just southwest of Anneliese's belly button, Leah listens, repositions the bell, and then listens again. I feel the frisson of waiting for confirmation. Then Leah smiles and, with her head still down, turns her eyes up to Anneliese. "Do you want to listen?" Anneliese nods, and Leah passes her the earpieces, and when the rhythm reaches Anneliese's ears, her smile spreads and inhabits her eyes. Amy listens also, and then it is my turn. Well, hello there, I think as soon as I hear it, tapping along over twice a second, and behind it the backbeat of Mom, bassier and running solid at half the speed. Then Leah takes my hands and guides me around the dome of Anneliese's belly, pressing my fingertips down as we traverse from spot to spot. "There's the head . . . that's the butt . . . the feet are here." As I do when feeling for something in the dark, I close my eyes for focus. "The head, the feet, and the butt," says Leah. "The constellation of baby."

Back around the little table we drink our tea and discuss everything from birthing tubs to doulas. Leah mentions that if Anneliese tests positive for group B strep, she will need an IV antibiotic, and since it can only be given by a licensed registered nurse, she would have to go to the hospital. "Well, *I've* got a nursing license," I say. "I can give it." "Perfect!" says Leah. In truth, my bravado is exceeding my confidence. Just what your wife wants during labor—you poking away at her veins trying to start your first IV in nearly twenty years.

The last thing is to set up the next appointment. When I say I can't commit to a date without consulting the calendar on my computer at home, Leah cuts a quick look at me and then at Anneliese. As we stand to say our good-byes, I look at the collage of photos on the wall—all babies Leah has delivered—but instead of the babies I am noticing all the beaming, gently hovering husbands and thinking I do not measure up.

But I leave the visit feeling better than when we came in the door. Above all it was good to see Anneliese so transformed. And I was impressed with the way Leah did her professional duties but allowed the visit to follow its natural course. She gave Anneliese all the time she required. There was no push to wrap things up, and not until we stepped into the hall and saw another family gathered did we realize there was an appointment following. During the drive home I do declare that if the term *herstory* is

used again, I will lodge a polite but firm objection. "False ety-
mology is no way to run a revolution," I say in the sort of declarative tones that precipitated the revolution in the first place. But
what really keeps circling my head was the phrase Leah used to
describe the landmarks of our child: *the constellation of baby*.
What a gorgeous image—the unborn infant afloat in the universe of mother, identifiable but unknowable.

On a cold night getting colder, we are off to fetch the guinea
pig. Amy knows we are going to visit Aunt Barbara and Uncle
John, but we haven't told her why. She has inklings, however,
and by the time we turn the final corner she has gone all wig-
gly. When we climb out of the van, she is in full pogo mode. In
the house, Barbara brings forth the guinea, and Amy pulls it to
her chest. "Oh!" she says, inclining her cheek to his. The guinea
pig nestles in. He has the coloring of a Guernsey cow, fawn and
white.

Now Barbara is loading John and me down with an astounding array of accoutrements—a ventilated travel box, Carrot
Crunchies, a bag of timothy hay, liquid vitamin C, claw clippers,
an exercise sphere, assorted rattle toys, two dishes, a gravity-drip
waterer, a cardboard crawl tube, and a fully stocked cage sufficient to house a brace of wombats. John and I lug everything out
to the van. As we pull out of the driveway, Amy is in her booster
seat, clutching the guinea pig to her heart. I look back to see her
face beneath the dome light, and with wide eyes she says, "I am
trem-buh-ling with *joy!*"

I turn back to face the road, so as not to dampen her happiness with my watering eyes.

By the time we're back at home in Fall Creek, it is bitterly below zero. I stoke the stove while Anneliese slices cheese, butters bread, makes a salad, and fries herself a burger. She is hungry all the time now. The three of us sit on the floor before the stove and turn the guinea pig loose. He roams, and we giggle at the sight of his bouncy behind, transported by two woefully undersize legs. Popular mythology holds that if the laws of aerodynamics are applied, bumblebees are calculably incapable of flight; watching his improbable giddyup, it strikes me that guinea pigs are the bumblebees of the rodent set. He toddles along like a wobbly furlined sausage, his butt all waddle and humpety-bump. He stays shy of Amy and me but makes regular loops back to Anneliese, putting his forepaws on her legs and begging food. At one point he tips her water glass and sticks his fat head inside, the scalloped surface giving him four red eyes. Amy beams through her missing front teeth. She is wearing her favorite purple footie PJs, and her neck and forehead still bear the fading red marks of the chicken pox.

Even with the fire going, the house feels chilly to me. I've been a little brittle lately. I think it's the move, the sick kid, the baby pending. For years I slid through life with no more on the line than my own hide. Now I have these other lives, and I'm feeling a little onerous, which is three syllables for whiny. I can't imagine how it was for my parents with everything on their plates. I look at the little girl so happy here, my wife with the baby in her belly, I feel the killing cold outside, and my head tumbles with the usual Big Questions, ranging from "Hello, God?" to "What character improvements are available via the adoption of a pet guinea pig?"

Amy wants the pig near this first night, so we allow her to unroll her sleeping bag beside the cage. Up in the bedroom then, I reach for Anneliese and hold my palm flat over the half-globe of the womb. No hiccups tonight. Palpating gently, I try to remember what I learned from the midwife. The constellation of baby. Even though it's dark, I close my eyes, straining to visualize what my fingertips feel. The head there, maybe? A shoulder here? I keep returning to one particularly prominent protuberance. It juts out, and I can't place it. I kiss Anneliese on the brow and roll over to sleep.

It seems we are bound to deliver a unicorn.

CHAPTER 2

I am building a glorious chicken coop in my mind. Each day I tweak the design based on an image I printed off the World Wide Web, or a weeded-up tumbledown model I spotted behind a barn while driving, or a photograph I found while paging through a 1928 issue of *Crows and Cackles* (edited by Prof. T. E. Quisenberry). I have sworn I will not house my chickens in blue tarps and chipboard. I am committed to providing them a sturdy, aesthetically pleasing home with a cute little drop-down gangplank, just like the one in my childhood edition of *The Little Red Hen.*

I don't understand architecture to any great depth, but some intangible element of the farm buildings of the early 1900s has always pulled at me. I can't put my finger on it any more than to say it has something to do with proportion, and they look like they *belonged* on the place. They weren't plastic or steel, they didn't look like oversize Fisher-Price accessories. I dream of a chicken coop that looks like it's been there a while. I want to view it from my window some hazy morning and imagine I am preparing to harness horses for the day's plowing. I will then grind a batch of imported free-trade coffee beans, fire up the computer, and update myself on the travails of decadent starlets. So it goes. Imagine the wizened quality of a life blanched of contradiction and double standard. And lest I disparage the Internet, today while noodling around it I wound up at the online home of the North Dakota State University Extension Service, where I discovered a historical archive of poultry housing plans dating back to 1924. It was like finding the Dead Sea Scrolls, without all the spelunking. We're talking Colony Cage Layer Barns, Broiler Breed Barns, Roll-A-Way Poultry Nests, Sloped Strawlofts, and—with an eye to the possible future—a Turkey Brood & Grow Barn. After three hours of close study and lost man-hours, I am currently mind-planning a gorgeous hybrid based on the Shed Roof Poultry House (particularly the 1933 and 1950 models) with significant modifications drawn from the 1951 Portable Brooder.

Vintage coop plans notwithstanding, history is not on my side. As a wannabe handyman, I am haunted by high hopes, false starts, and even worse finishes. Evidence surrounds me: the engine heater I bought and left beneath the truck seat until the packaging fell away; the bathroom faucet I bought after purchasing my New Auburn house twelve years ago and never installed (it's still there, under the sink); a perfectly hung screen door (I hired a guy). I can trace the trouble all the way back to when I was five years old. Dad was remodeling the barn, and I decided to help. It was deepest winter, but I bundled up and trekked outside, determined to pitch in. Dad handed me a hammer. First thing I did was lick it. Rather than the sweet electric taste of the shiny steel, I felt a numb, crinkly sensation as the hammerhead froze fast to my tongue. Panicking, I yanked it loose, pulling away a perfect circle of skin. I forsook carpentering and went into the house to read comics and taste the raw spot over and over.

One does not become a farmer simply by taking possession of a milk cow, but it does drag you in that direction. The night Dad tied that Holstein to the Falcon, he tied an anchor to his ankle. From that day forward he would find his way to the barn a minimum of twice a day, *every* day, morning and night, seven days a week, with no break, year after year after year. Whenever we went to Christmas dinner, or visiting of a Sunday afternoon, Dad kept shooting looks at the clock. Sometime around 4:00 p.m., he'd say, "Weeelll, I s'pose . . ." Then someone else would push back from the table and say, "Yep, them cow's ain't gonna milk themselves," and we headed home.

Dad went pretty easy on us with the milking. While many of our schoolmates had to milk every morning and night, John and I traded off helping Dad every other evening. The chores began when Dad switched on a vacuum pump plumbed to pipes that ran the length of the barn and created the negative pressure that drew the milk from the cow's udder into the pail of the milking machine. The pump was vented outside via an exhaust pipe that emitted a noise similar to an extended bout of oily flatulence. If I heard the vacuum pump start before I crossed the yard after supper, I knew I was running late.

Before applying the milker, we washed each cow's udder with a cloth and warm soapy water. This removed any caked dirt or manure, but it also stimulated her to let her milk down. By the time you returned with the milker, beads of milk were forming at the end of each teat. Our De Laval Milkers were composed of a stainless steel bucket that sat flat on the floor and was capped with a detachable top sprouting several sets of hoses. One set was plugged into the overhead vacuum pipe. The other two hoses—a narrow black "pulse tube" to provide vacuum, and a larger clear tube to carry the milk—were connected to a shiny silver claw from which radiated four hollow rubber tubes called inflations. The inflations were collared by individual stainless steel shells that created a potential space wherein the air pressure was alternately lowered and released by means of a revolving mercury switch and a wonderfully named unit called the pulsator. The pressure changes drew the milk from the teats through the inflations. Almost immediately after you opened the vacuum valve and swung the inflations into place, a white trickle appeared in the clear tube, and shortly a steady wash of milk was pulsing into the pail. If the cow was a good long milker, you had time to go scrape cow pies into the gutter or perform some other small chore before she was ready to have the milker removed. Sometimes you'd hear a *suck-whooosh!* followed by a muted clatter, and when you got back to the stall the milker had been kicked to the cement and was vacuuming straw chaff while the cow flicked her ears in irritation.

If the cow was a regular kicker, I'd stick right with her, pressing the top of my head into her flank and weaving my left arm in front of her near leg and behind the udder to grip the large tendon just above the knee of the far leg. When the cow raised her near leg to kick, I pressed my head in hard, hung on tight, and raised my left shoulder to keep her from getting a hoof over.

Usually after about three or four rounds, she'd give up. But now and then you got a cow heavily into bovine karate, and you'd have to take additional measures. Some farmers used a piece of binder twine to tie the cow's tail to a nail driven into a beam overhead. Others had someone else hold the cow's tail in a twist until she milked out. Some farmers used hobbles. Others designed medieval anti-kicking devices. We once had a cow who treated every milking as an audition for the Rockettes. We tried everything I mentioned above plus a few other tricks. Nothing worked. Night after night she peeled the milker off and stomped it to the straw, occasionally nailing one of us in the thigh, or flicking our ears with her dewclaws. Dad must have been talking about her at the feed mill, because some farmer lent him a homemade anti-kicking tool, which was basically a giant horseshoe-shaped clamp that fit around what you might imagine as the cow's waist. The clamp was held in place by means of a screwjack apparatus.

Dad put the clamp in place and snugged it down. Then he attached the milker, and the cow kicked it off immediately. He tightened the clamp a few more twists. She kicked the milker off again. I have never heard Dad cuss, but he must have been close. I will say that by the third time he gathered up the milker unit, he had developed a firm set to his jaw. Taking a deep breath, Dad gave the handle two more vigorous twists, at which point the cow collapsed in a heap atop the milker. The cow was fine, if flustered, but it took us the better part of ten minutes to tug that milker out from beneath her.

Dad hung the clamp on the wall. It dangled there for a few months until the man came to pick it up, and we never did get that cow to stop kicking.

When the milk thinned out and streaked through the clear main hose like white raindrops wisping across a windshield, you cut the vacuum by twisting a petcock and lugged the suddenly heavy milker to the walk. Occasionally a cow topped the pail, but Dad never went crazy pursuing production. He'd add a little salt and mineral to the cow feed, but beyond that, he said if we don't grow it, they don't eat it. Some of our cows produced a hundred pounds of milk per day, but they'd have to drop below forty pounds before Dad sent them down the road.

In the beginning we shipped our milk in cans. The cans stood about three feet tall and weighed 100 pounds when full. I remember the warmth of them full with fresh milk inside the tiny stand-alone milk house, then Dad lowering the cans neck-deep into a concrete tank of cool water hidden beneath a heavy wooden lid. When Dad refreshed the water, the overflow traveled through a pipe to the stock tank in the barnyard. The pipe ran about three feet above the ground and was polished smooth; we often used it for gymnastics and tightrope walking. Every other day Gene the milkman would pull up with his cavernous steel truck and, in a rattle-bang exchange, trade us empty cans for full. Gene was sinew-skinny, but he kneed those full cans aboard the elevated truck deck like they were packed with cotton.

In order to get a better rate for his milk, Dad eventually built a new milk house. This one was filled with fun doodads—a double stainless steel sink, liquid soap dispensers, a paper towel roller just like the one in public restrooms (it was required to make inspection; to save money, Dad never actually let us use any of the paper), a double-sided trapdoor in the wall to admit the hose that pumped the milk to the truck (we used the trapdoor to play mailman), a door that swung both ways between the barn and the milk house (it reminded us of the ones we saw in restaurants), and a shiny stainless-steel bulk tank the size of a hot tub. We loved to pull the tiny cymbal-like lids and watch the revolving paddle swirl the milk as it cooled, and on hot days we would go around the back of the tank and stick our hands elbow deep in the water reservoir to touch the cooling coils when they were fat and silky with ice.

During construction, while the walls had yet to be enclosed, my sister Suzy was playing farmer, and my brother Jed was her cow. In need of a stanchion, Suzy had Jed stick his head in the gap between two studs. Later we were all seated at the dinner table when Mom noticed Jed was missing. "Oh," said Suzy nonchalantly, "he's in the milk house." And so we found him, on hands and knees with his head jammed between a pair of two-by-fours. Having pushed his way in, he couldn't back out. Dad levered the studs apart and freed him. I was always confused when city kids asked us how we had fun without a television.

Dad ran two milking units with three buckets. We dumped the spare bucket while two more cows were being milked. He built the milk-house floor four feet lower than that of the barn, so when we stepped through the swinging door, we stood on a concrete landing with our feet at the same level as the bulk tank. Rather than having to lift and dump the milk, we just stepped across the gap, put one foot on the corner of the tank, and dumped from ankle height. The milk passed through a strainer, draining slowly away to leave a round cap of snow-white froth atop the disposable paper filter. Toward the end of milking the barn cats would start hanging around the door, waiting for the moment we tossed the filter in the gutter and they could lick it clean.

While Dad milked the final cow, I fed the calves. Originally Dad used a powdered milk replacer we mixed up in a bucket, but over time he took to putting air quotes around the word "replacer" and went back to the real stuff. The calf bottles were roughly the size of a milk carton and capped with a rubber nipple. The calves wriggled their tails and sucked with such frantic exuberance that if you didn't hold the nipple tightly, it burst off and the milk hit the concrete in a big white splash. You also learned not to stand directly behind the bottle, as calves have an innate tendency to pause in their suckling in order to deliver a vicious head butt originally intended to get Ma to let her milk down. If you're a little kid with a bottle in your solar plexus, the only thing let down is tears. If you're a little boy holding the bottle even lower, you find yourself gasping wordlessly in the gutter. The bottle was empty when you heard the *squinch-squinch* of the calf sucking air. You had to give a mighty tug to get the nipple free, and the calf didn't want to stop. Sometimes we let them nurse our fingers, and I remember warm slobber, the ridged roof of the mouth, and the scratchy scrub of the tongue.

When you heard the vacuum pump power down, you knew Dad had thrown the switch, and the milking was done. In summer we'd walk the mangers, unsnapping each cow from her neck chain. They'd back half out of their stalls and, like a tugboat in an undersize boat slip, make a ponderous turn followed by a teeter-totter lunge sufficient to clear the gutter, then head for the door. They'd go to the pasture, and we'd go to the house.

I got religion in the third grade, and jeepers, did I need it. The devil was in me, and Hardy Biesterveld wasn't helping. Through second grade, I had been a precocious model of good behavior, with fine marks and a talent for reading two grades ahead. Hardy was a year ahead of me, but then he was held back a year, and when I crossed the hall we became classmates.

I do not know what might have spawned my recalcitrance—I recall no particular psychological trigger point other than the fact that Mrs. Zipstrow, the second-grade teacher, was known to fling staplers—but one day there I was, sitting in the hall with Hardy Biesterveld, taking turns to see who could string together the longest unbroken run of swear words. We had been banished to the hallway for disruptive behavior, and indeed, we had taken to hanging out in the classroom as if it were a street corner. Mrs. Kramschuster was in her rookie year, and we threw sand in her gears at every opportunity. Hardy once proposed that we conceive of her as a ball of rubber cement. Holding one hand about a foot from our eyeballs, we pinched her image between our thumb and forefinger, mimed rolling her between our palms until she was the size of a marble, and then played catch with her. We bounced her off the blackboard, and we stuck her to the ceiling. When she told us to behave, we generally complied, but smirked, rolled our eyes, and insulted her beneath our breath. We sometimes cast her in the composition of indecent rhymes. We were a pair of impudent slyboots.

My moral decline was exacerbated by Mrs. Lovelace, a teacher's aide. *Bountiful* is the only word that will do. Young, blond, and newlywed, she elicited within us a naive friskiness. One day she came through the classroom door clad in a black velveteen top with a neckline that appeared to have been cut around the prow of a galleon. Suddenly we were eager swots in search of constant tutoring. Again and again we lugged our geography workbooks to her table. Mimicking Hardy, I leered knowingly, but inside I was trembly with prepubescent wonderment. That profound mammate cleft, framed in a breathtaking swoop of embroidered décolletage—the vision pressed itself warmly into my young brain. Hardy and I marked the occasion by composing crude boob jokes, at least one of which incorporated an internal rhyme scheme. I was being drawn down the path of wickedness. Here is Mrs. Kramschuster, writing in my first-quarter report card, under section II, Student Attitude to Date: "Mike is a polite boy in class and displays a mature ability to get along with his classmates. Cooperative and responsive. . . . He is prone to visiting recently, which is affecting his studies and progress."

Mrs. Kramschuster, second quarter: "Mike achieves more success when not distracted by Hardy Biesterveld. Mike is becoming more sensitive to others but I am afraid his friendship with Hardy may affect this. . . . Appears to have developed better self-control, with the elimination of moodiness."

Mrs. Kramschuster, third quarter: "Continues to waste time. Mike appears belligerent when asked to get to work. . . . Appears more moody."

Sounds like a boy who needs to get right with Jesus.

The 1936 edition of *The Best Loved Poems of the American People*, selected by Hazel Felleman, is a 670-page brick. Ms. Felleman, a longtime editor of the Queries and Answers page of the *New York*

Times Book Review, is hailed in the introduction by Edward Frank Allen as "the liaison officer who has coordinated the poetry preferences of the nation." Our copy resided on a shelf beside the Monarch woodstove. *Best Loved* is arranged in twelve sections. I spent the majority of my time in section VIII, "Humor and Whimsey." "The Animal Fair" and "How Paddy Stole the Rope" were favorites. But one night I stopped off in section III, "Poems That Tell a Story." And on page 229, I came to a poem titled "The Hell-Bound Train." It scared the bejesus out of me.

It has been twenty years since I read "The Hell-Bound Train," and over deer hunting season this year, when I was in my parents' house, I had another look at it. I couldn't remember much about the poem, just the idea of a locomotive steaming for hell. I recalled an image of the devil stoking the steam furnace.

Turns out the poem's main character is a cowboy. I had forgotten that:

A Texas cowboy lay down on a barroom floor
Having drunk so much he could drink no more;
So he fell asleep with a troubled brain
To dream that he rode on a hell-bound train.

I remembered none of this. But there, in the second stanza, was the image that had scared me silly:

The engine with murderous blood was damp
And was brilliantly lit with a brimstone lamp;
An imp, for fuel, was shoveling bones,
While the furnace rang with a thousand groans.

The lines hit my third-grade gut like an electric acid ball. Reading the next ten stanzas was like walking through a house of horror—the lost souls "all chained together," the air becoming "hotter and hotter" until "the clothes were burned from each quivering frame." There was shrieking and begging, there was the devil, capering and dancing for glee. "You have . . . mocked at God in your hell-born pride . . . so I'll land you safe in the lake of fire . . . where your flesh will waste in the flames that roar."

In the last two stanzas, the cowboy startles and wakens with an anguished cry. In great desperation, he prays for salvation.

And his prayers and his vows were not in vain
For he never rode the hell-bound train.

I ran to the bathroom.

I stood between the toilet and the sink, teary with fear, praying that I might escape the hell-bound train. I stood there a long time. When I had finally composed myself, I cut quickly through the light of the dining room, up the stairs, and straight to bed. Dad was at the kitchen table, but I didn't want to talk. Beneath my quilt and with quaking heart, I promised God I would do better. At some point in the supplication, I slept.

When I checked out a copy of *The Writer's Market* from the L. E. Phillips Memorial Library sometime in the late 1980s, I was gainfully employed as a registered nurse and had not the slightest conception that I might one day actually pay the rent by writing. I've been surviving at it in one form or another ever since, and so far so good, but as a freelance operator you never get clear of the sense that the current gig might be your last, and this is what drives me to hunt down magazine assignments and pitch the next book and hit the road for days on end with boxes of books in the trunk and my well-worn anecdotes at hand. We're in good shape right now (it helps when I get my annual Social Security statement and see all those four-figure years in the not-so-distant past). Still, as I once heard someone put it, the secret to successful self-employment is to wake up scared every morning, and I usually do.

In public, I am prone to saying freelance writing is a slightly less reliable way of making a living than farming. But when I think of all the hungry mouths circling the dinner table, and Dad out there in the barn trying to make a living with the milk of eighteen cows, I scale down the drama. Dad supplemented the milk check with some on-call work for a factory in Bloomer over the years, and Mom earned money from the county for providing foster care, but even with that in the equation, the family still qualified for government cheese. For years Mom refused the cheese, in part because she felt that she and my father—with their education and opportunities—had made what we now call "a lifestyle choice" and shouldn't expect any help towing the barge. But when my sister Rya—dying of congenital heart and lung failure, her potassium levels depleted by diuretics—got to where she would drink only orange juice, Mom followed the urging of a county social worker and signed up for the program so she could buy OJ concentrate by the case. She did the same thing when my little brother Eric, on gastric tube feedings from infancy until the time of his death at ten, needed special formula. And sometimes the social workers just insisted that we take the cheese. I remember it in the fridge, pale yellow inside the cardboard box, more like a giant pencil eraser than cheese.

We grew up poor but not wanting. Our clothes were nearly all secondhand and came packed in cardboard boxes, but I remember the arrival of the boxes as an occasion for excitement rather than shame. We clustered around as Mom sorted the booty, hoping somewhere in there was a cool T-shirt our size. When we pulled the shirts over our heads, they were already broken in, but the scent of unfamiliar detergent made them seem new. Once a year at the end of summer we went to Chippewa Falls to buy school shoes from a store owned by a man named Ed, who kept seconds and overstocks in the back for families just like ours. When we got home with our shoes, we'd bail out of the car and rip around the yard, convinced that this year's tennies were the speediest ever. "These have good treads," I'd say, cutting sharply like a running back.

Mom pinched pennies wherever possible, perhaps nowhere more than in the area of food. But we never went to bed hungry. Back before the big-box buying clubs of the present, there was a dingy warehouse in Eau Claire with great racks of off-brand and damaged goods. Rarely did we open a can of beans that didn't look like it had been backed over by the truck that delivered it. And the Great Generic Craze that struck in the mid-1970s was right up Mom's alley. The pantry was stacked with white cans labeled BEANS and white bags marked MACARONI. I am not a picky fellow, but I am not opening a white can that says TUNA. In later years, after I left the farm, my brothers provided harrowing details of meals comprised entirely of frozen leftovers from the county jail, but I am not going to pursue this story, as I'm not sure who Mom knew on the inside, and I'd hate to land her smack in the middle of a scandal at this late stage.

The savings began with breakfast. Five days a week we got oatmeal. Plain, gray, factory-grade oatmeal possibly useful as masonry mortar. In fairness to Mom, there were occasional deviations into decadence—farina with raisins! cornmeal with molasses!—and on Fridays she would indulge her profligate inner hedonist by stirring fourteen generic chocolate chips into a pot of oatmeal the size of your head. (In fairness to Mom, she has recently produced a breakfast schedule written on a recipe card that seems to indicate there was less oatmeal than I remember, but for all I know she made that card up a week ago, and I'm sticking with my story the way that oatmeal stuck to my ribs.) During a particularly impecunious stretch we cut back on oatmeal and ate boiled wheat, because the neighboring farmer let Mom scoop it straight from the grain wagon into a garbage can, thus qualifying her for the wholesale rate. Some of it we ground into flour. On Saturday we got pancakes, but I was legal to vote before I realized you do not make maple syrup by dissolving two tablespoons of brown sugar in a pan of hot water. Bottom line: if you had breakfast at our house on a weekday, odds are it originated from a twenty-five-pound bag or a thirty-two-gallon plastic trash can.

But Sunday . . . on Sunday we ate cereal from a *box*. *Boughten* cereal, we called it. Cereal like all the other kids ate. All the other kids, that is, whose mothers bought discounted off-brand raisin bran in damaged packaging and never went shopping until she had attacked the *Chetek Alert* coupon section like D'Artagnan on a bender. When she finally paused to let her scissors cool, the newspaper looked as though it had been caught in the cross fire of a street fight conducted with X-Acto knives and a confetti cannon. Upshot being, if we got cereal, it was either on sale, a two-for-one special, or one of those "replica" knockoffs in a plastic bag (so much for cereal in a box). How I longed for Apple Jacks. Still, on Sunday there was no oatmeal, and that was enough.

Sunday's "boughten" cereal represented a rare concession to expedience. When you are struggling to get six or eight kids ready to leave for church by 9:30 a.m., and some of those kids are developmentally disabled or run-of-the-mill recalcitrant grumps (Your Author), you are happy and wise to line up the cereal boxes like books on a shelf, toss out spoons and a stack of bowls, and let the rug rats have at 'er. And man, we could get through some grain. One time we were watching television at Grandma's house and a Total cereal ad came on, the one where comically stacked bowls of competitor's cereal are juxtaposed with a single bowl of Total cereal, and the announcer says something along the lines of, "You'd have to eat *fourteen bowls* of Frooty-Os to equal the nutrition in *one* bowl of Total," at which point my brother John turned to me with resolve and said, "I'll take the fourteen bowls."

Ours was not a loud family, but once Sunday morning breakfast got rolling, the action was steady, with the sifting sound of cereal sliding from the wax-paper lined boxes and the high-tension ping of spoons against the rim of Corelle Ware dishes. I loved to look at the pictures on the boxes while I ate, and dream of the day I would save enough box tops to get a real jet airplane. When at an early age I began to learn to sound out words, my Sunday morning cereal time was the source of great strides in reading comprehension. I'd read the boxes side upon side. By the time I was in kindergarten I could spell *niacin* and *riboflavin* with dispatch. In my sullen years, I would arrange three cereal boxes in the manner of a cubicle and seclude myself for a good read.

If I can continue to support my growing family on a freelancer's wages, I will have my wife to thank. Early in our courtship Anneliese picked me up for a date in a battered Honda and apologized, saying she was too cheap to spend money on a new car. Unbeknownst to her, I found this comment the equivalent of a red satin nightie. Although she eventually sold the car and upgraded to our current $1,000 van, her frugality remains constant. In the shower today I bumped into a gigantic thirty-two-ounce bottle of shampoo (as a middle-aged bald man walking, I find thirty-two ounces of shampoo to be profligate in the extreme). I also noticed that the special unsecret ingredient in this shampoo is *placenta*. Anneliese goes in for some alternative concoctions, but even so this seemed a bit much. Upon closer inspection I saw there were several price restickerings on the bottle, with a final markdown to $2.29, and then I understood. Still, I skipped the placenta extract and went with a mini-bottle I scored from a Super 8 outside Wichita. Free, and it put a fine sheen on my scalp.

Matrimonially speaking, being of one mind on money matters really smooths the sheets. While other spouses go ballistic over nasty surprises on the credit card, I am reduced to waving the Visa bill and barking, "$7.95 at Goodwill?!? That's the second time this *year*!" When the winter winds whistling through our overworn upstairs windows forced us to turn on the baseboards and spiked the electric bill, Anneliese talked my contractor cousin out of a couple sheets of pink Styrofoam and trimmed them down to fit. Now our entire upstairs is bathed in the soft pink glow of love and free insulation.

Like most well-worn tropes, the idea that a man looks to marry a dead ringer for dear old Mom is probably only half accurate at best, but when I go to the kitchen sink and find plastic bread bags air-drying on the faucet handles (the twist ties are stored neatly in the drawer beside the repurposed plastic picnicware), I will admit to the ol' déjà vu. Right this moment there are dented cans of off-brand black beans in the pantry, and I recently came through the front door to find my way impeded by a twenty-five-pound bag of garbanzo beans. I am not kidding, and apparently my future holds a lot of hummus.

Once a man came to load one of our cows for the sale barn, and before Dad could get out there, the man had whipped it until there was blood on its back. Before the man was out of the yard, Dad was on the phone to the shipper. Don't ever send that man again, he said. Another cattle jockey came in the barn carrying an electric cattle prod. "You won't need that," Dad said. The man said something about how good it worked. "You won't need that," Dad said, a little more deliberately this time, and the man returned the prod to the truck. If a cow was being stubborn, we were allowed to smack her on the flank with an open hand, but that was more for the sound effect than anything. We could also tap them with a broom handle on the spinal ridge where the tail attached, or twist the tail—although often as not the tail-twist made them slam on the brakes.

I recall striking only one cow in anger. All told, I tried to hit her three times, but the last time I whiffed. Her name was Belinda, and she was a "rooter." If you turned your back on her while cleaning the manger, she'd "root" you from behind, jamming the rock-hard brow ridge of her skull under your coccyx and boosting you headfirst into the wall. Sometimes it hurt and sometimes it didn't, but it always got your full undivided attention, and it consistently tripped my rage trigger. Once I was busting hay bales, and she rooted me right off my feet. I whipped around, balled up a fist, and punched her right between the eyes, hard as I could.

Are you familiar with the real estate between a cow's eyeballs? For the purposes of simulation, drape a thin rug over a concrete block and then hit it bare-fisted as hard as you can. The vibrations reached clear up to my ears, and the numbness persisted for twenty minutes. As I huddled against the wall, cradling my useless arm and wondering how best to splint it, the cow regarded me placidly. My best pile driver, and it had less effect than the touchdown of an anemic horsefly. The next time she tagged me, I was sweeping up the manger. Wiser now, I whipped around and smacked her over the skull with the broom handle. Same net effect—she just blinked at me—but more trouble, because the handle snapped, and I'd have to explain that to Dad. Later when he asked me why I used a whole roll of black electrician's tape on the broom handle I told him Belinda knocked me over and I *fell* on the broom handle. I think he knew, and just let it go, because that cow was flat crazy. Some cows would take a shot at you now and then, but she was one of the rare ones who would actually come after you. One summer evening all the cows came in for milking except Belinda. I grabbed the big rubber mallet Dad used to knock the feed loose from the side of the bin and went out looking for her. Rather than run off when she saw me, she waited until I got near, lowered her head to freight-train position, and came thundering at me.

For the first minute or so, I fared pretty well. I'd run in a straight line until I could feel the thud of her hooves, then I'd cut a real tight turn. While she slowed down to change directions, I sprinted clear again. With every juke I kept trying to work my way closer to the fence and safety, and before long we had zig-zagged our way to within about twenty yards of the woven wire, but I was getting winded, and that cow hadn't lost a step. Finally, when I cut two corners not quite tight enough and she tagged me with a half-root, I realized I had to make a break for it. I still had the rubber mallet, but if I squared off to whack her, I risked getting trampled. Instead, I decided to fling it at her head in the manner of throwing an ax, hoping to clock her good enough to slow her down. Gripping the mallet handle tightly and running full tilt, I looked back over my left-hand shoulder, gauged the distance and, still on the run, pivoted halfway around and flung the mallet at her crazy-cow noggin with every bit of strength I could muster.

And missed her completely.

Oh, my goodness, I remember thinking.

Sure now that she was emitting cartoon smoke from both nostrils, I made one last valiant sprint straight for the fence. She was at my heels and gaining when I launched into a full-out dive. Grazing the top row of barbed wire, I performed a credible tuck-and-roll and hit the soft ground on the other side. After a nice little rest, I went off to find Dad, and not long after that, Belinda went to market.

Back in the day, most farmers kept a bull on the farm for the obvious purpose. We all knew a few stories of goring, trampling, and death. What Dad had instead was a cabinet mounted just inside the milk-house door. The cabinet door—which folded down to serve as a miniature desk—was imprinted with a silhouette of a fine bull, the words EVERY SIRE PROVEN GREAT, and the logo ABS, for American Breeders Service. Within the cabinet were a few stubby pencils, a few bright tags that read BREED THIS COW, and the American Breeders Service bull catalog.

The ABS catalog was basically *Playgirl* for cows. It was filled with page after page of photographs of the ultimate bulls. These were the Greek gods of the bovine world. They were posed with their front hooves on a small mound of clean sawdust, and their tails hung long and were fluffed to a voluminous switch. The bulls were ornately named. One of the stars of my childhood was Fultonway Ivanhoe Belshazzar—one-third landed gentry, one-third literature, and one-third Old Testament. I always thought it would be fun to be the guy coming up with names for the bulls. I figure you'd want something relevant but exotic, say, Golden Turkish Alfalfa Rocket.

When a cow was in heat (we learned early to listen for the urgent, high-pitched mooing and cows "riding" each other), we kids would go through the catalog page by page, studying each portrait closely. In addition to the photographs, each bull's page included a chart delineating their specific genetic attributes relevant to the qualities they caused to arise in their female offspring—which, after all, was where the farmer's prime interest lay. Among the categories you might review were body depth, foot angle, thurl width, rump angle, teat placement, and udder cleft. We'd pore over the photographs, review all the data, and then finally pick our favorite. Dad, we'd say, this one here— Spanky Tango Cremora Blaster—he's the one!

Knowing now what I didn't know then about my parents' financial situation, I have come to realize Dad probably just went to the back of the catalog, to the discount section ("Bull in a Bucket"), and ordered the cheapest product available. And then, sometime within the next eight hours, the artificial inseminator would arrive, and he would walk into the barn and commit astounding acts.

When you're a kid growing up on a rural Wisconsin dairy farm with no television, the artificial inseminator is a combination science exhibit and freak show on wheels.

We never missed it.

The inseminator (we called him "the breeder man") would roll into the yard in his pickup truck, and in the back he would have this stainless steel canister about the size of a beer pony. The canister was filled with liquid nitrogen, which kept the semen frozen at −321 degrees Fahrenheit. The ampoules were suspended on a rack. When he popped the lid on the canister, mysterious wisps of fog would boil up and spill down the sides, evaporating halfway down to the truck bed. Sometimes he would allow us to dip a length of string into the nitrogen. When we pulled it out, it was frozen solid and could be snapped like a twig.

After extracting the semen, the inseminator placed it into a short syringe, which he then attached to a long, slender pipette.

Next—and I'm not sure if this was standard procedure, or just our guy's particular personal flair—he would place the pipette crossways in his mouth and grip it in his teeth in a sort of grimace. I remember this very clearly because we would be waiting inside the barn on the walkway and the inseminator would step through the barn door all backlit by the sun, and he would be wearing those tall rubber boots and holding that straw in his teeth, and I was always reminded of a pirate boarding a ship.

I assume the cows had a similar reaction.

Dad would hang a paper tag from the rafter behind the cow he wanted serviced. After locating the tag, the inseminator stopped behind the cow, drew on a shoulder-length plastic glove, and stepped across the gutter. After patting the cow to calm her, he grabbed her tail, hoisted it, and from then on the whole deal was very personal.

I can't say the cows ever appeared overly distressed by what certainly had to be a disruption in their day. They would pause in chewing their cud, kinda freezing in a "hunh?" sorta pose, and their eyes would bulge a tad, about like yours would at the point of realizing your taxes were due yesterday. Occasionally one would engage in a little do-si-do (who wouldn't?), but all things considered, their reaction to having a stranger's arm elbow-deep up the rectum was positively restrained.

I have met a great number of artificial inseminators over the years, and they are nearly always cheery about their profession. Apparently a career spent operating at less than arm's length from the place where the miracle of life and its base by-products intersect engenders a certain jocular pragmatism. One of our inseminators was pleasant enough, but at the feed mill there were rumors of his drinking. Perhaps so, said Dad, who abhorred alcohol in all its forms. But we had also just come through a stretch in which the allegedly drunken inseminator settled twenty-four cows on the first try, and twenty-three of those cows had heifer calves. If that man *was* drinking, Dad said, paraphrasing the apocrypha of Lincoln on Grant, we better find out *what* and get him some more.

We observe our heroes and emulate accordingly. When my brother Jed was still in training pants, Mom found him with his arm wrapped in a plastic bread bag and jammed inside a roll of butcher paper. He had a green Tinkertoy rod crossways in his teeth and was patting the butcher paper to calm it before delivering the coup de grâce.

There are chicken books in the bathroom, *Backyard Poultry* clippings on the bedside stand, and coop sketches scattered around my desk. Anneliese is in the spirit as well, quoting from *Chickens: Tending a Small-Scale Flock for Pleasure and Profit* and referencing the chicken tractors of Joel Salatin. But I am also prone to nattering on about where we'll put the pigs, and how maybe we should fence off a patch for a pair of beef cows, and how I read in *Countryside & Small Stock Journal* that goat meat is gaining popularity, and also wouldn't it be terrific to fence the yard for sheep and save the gas money? I know I said at the outset all I wanted was some eggs and perhaps a slice of homegrown ham, but here we are with thirty-seven fallow acres. . . .

I keep trying to rein myself in. It's not far from champing at the bit and biting off more than you can chew. We have a smallish tractor here on the farm, and yesterday the battery went dead.

No problem. I pulled the pickup truck beside it, hooked up the jumper cables, and—rather than rev the engine impatiently—went off to multitask while the battery charged. When I returned ten minutes later, the interior of the shed was a haze of toxic smoke and the battery was fizzing like a junior high science project. There are only two ways to hook up a battery—the right way and the wrong way—and the right way is *color coded*. So now I had to replace the battery. I couldn't find the correct wrench, and the one matching socket I located was stripped. That meant I had to pry the battery loose using cheap vise grips and a screwdriver. The cold morning air rang with curses.

I finally wrestled the battery loose and set off to trade it for a new one at Farm & Fleet. While there, I noticed a bin of cheap wrenches. No self-respecting handyman buys cheap wrenches, so naturally, I was interested. The wrench sets were in two separate bins, but the price was the same, so I just grabbed the nearest set. Back home, I was almost giddy at the idea of installing the battery now that I had the proper tools. I unrolled the bag of wrenches to select a half-inch, only to find every wrench marked with "mm" instead of "inches." Two bins of wrenches, and I managed to pick the *metric*. The battery bolts (and pretty much everything else on the farm) are *standard American*.

Good news is, you can fling a metric wrench forty feet, no conversion necessary.

The dead tractor battery reinforces what experience has taught me over and over again: Don't overreach, farmer boy. It will be miracle enough if I can build a coop that will keep my chickens dry. Tonight as I stand and watch the sun go down above our barren spread, I am reminded that for all my talk and bathroom reading, what we've got here so far is thirty-seven snowbound acres and a guinea pig.

I don't know if I was born again the night I read "The Hell-Bound Train." Three years would pass before I professed my faith before members of our church. But there in the bathroom that night, that was my come-to-Jesus moment. This was when it hit me that any little boy who hung out cussing with Hardy Biesterveld would never breach the Pearly Gates. For the first night in my sheltered life I desperately craved sanctuary—from the cackling devil and his hellfire coals, sure, but also from myself. From the filth of my own weakness. I don't recall, but I can't imagine I strolled into school the next morning and told Hardy Biesterveld I was swearing off swearing. I do think I quit the cussing cold turkey, but as far as the rest of my scampitude, I reckon I just scaled back gradually. Didn't dig my heels in, but dragged my feet some. I know we stayed on friendly terms right into adulthood. I just didn't follow everywhere he led. And I'm glad I didn't write him off. In the first place, that would have been snotty. In the second place, as the decades have unfolded, I have found great wisdom in the company of sinners—wisdom not always available via pristine living. And as a guy who equates sin with furtiveness and great lashings of guilt, I have always felt a certain awe for sinners who lay it all out there full-force.

On the twenty-fourth of May, 1974, I received a photocopied diploma affixed to a piece of green construction paper. Mrs. Kramschuster joined the two pieces of paper using rubber cement, and three decades later I can see the brush-swipe patterns where the cement seeped through, and the memories come flooding back. How the rubber cement swabbed on your skin with an evaporative coolness and slick like snot, but if you rubbed it together it dried out and became rubber, much as wet snot rubbed between the palms of your hands will become a serviceable booger. It reminds me of Hardy Biesterveld and how we would slobber rubber cement on our palm and then rub them together until we made our own off-kilter superballs. How we'd sniff the open bottle, the fumes putting a cool burn in our nostrils. And of course it reminds me of how we treated Mrs. Kramschuster. "THIS CERTIFIES," the fancy script says, "Perry, Michael has completed the studies prescribed for the 3rd grade and is hereby promoted to the 4th grade."

Mrs. Kramschuster's signature is Palmer-penmanship neat. We can imagine her relief.

In the winter, darkness fell well before supper. By the time I followed Dad out for the evening's milking, Orion was climbing from his kiva in the woodlot behind the barn, and the clear night air was tin-pail cold against my nose. The barn windows glowed an opaque yellow, and during the walk I anticipated the bare-bulb interior, bright with all the naked incandescence reflecting off the whitewashed walls and rafters. When I pushed through the milk-house door and into the light, the warmth—a thick sachet of alfalfa and manure—rolled around me with such fullness I felt I could tug it to my shoulders like a quilt.

Eighteen Holsteins and a passel of calves easily generate a barnload of body heat, especially when it's all concentrated beneath a low ceiling insulated by hay bales stacked twenty feet deep. Sometimes during the day when the cows were settled we kids went to the barn and lay lengthwise along the backs of the tamer animals to absorb their warmth. Because of the way she tucks her hindquarters, a cow at rest tilts off-kilter, allowing you to nestle rump to withers against the ridge of the backbone while draping your limbs across a hemisphere of abdomen. You rise and fall with each bovine breath, and if you hold especially still you will feel the subterranean thump of a five-pound heart. At regular intervals the cow will lurch softly and summon a cud. The dewlap ripples, and a wad of ruminated forage rises visibly up the throat. Rolling the bolus to her tongue, she'll work her jaw forty or so times, swallow, wait a patient moment, then raise another. It's hard to imagine regurgitation as a form of meditation, but for cows, it is so. If you feel the animal rock forward, it is time to bail. She is working up the momentum to rise, and it is critical to get clear before she heaves to her feet, hooves scrabbling on the concrete or perhaps your toes.

Barn cats also covet the warmth of cows, but their approach was the reverse of ours; they waited until a cow stood, at which point the cat cruised in to curl up in the warm straw where the cow's belly had rested. Fine and dandy, until the cow decided to lie down again. A cow does not lower itself gently to earth but rather shuffles about a bit and then pulls the rip cord. The older cats were usually wise to this, but now and then a youngster got caught. When a cow parks on a cat, the cat shape-shifts. In short, there is an increase in square footage. We called them pancake kitties.

All those years ago I already knew I didn't want to milk cows for a living, and yet those winter nights in that barn remain in my memory as sanctuary. I can see Dad down on one knee, head bent to the black-and-white flank, watching the milk course through the clear tube from the udder to the pail. The very provision of his family, passing before his eyes. I was a kid, so it never occurred to me to wonder what was in his head—if he was running the math on this month's groceries, or preparing a ruling regarding the latest bad news from school, or just longing for a full night's sleep—but I absorbed a deep reassurance from his posture. Once when I was still a small grade-schooler but old enough to help with the chores, a traveling salesman drove into the yard, jumped from the car, strode up too close, and patted me on the head. I remember his knee bouncing behind creased polyester pants, and then, hearing the sound of the vacuum pump, he said, "Where's your daddy? Pullin' tits?" "He's *milking cows*," I said, as coldly as a grade-schooler could, quietly furious that this shiny-shoed stranger would barge up our driveway and profane my father's work.

Lest I create the impression that every milking session was a hushed ritual of patriarchal ceremony, I should add that between cows Dad dangled upside down from a bar bolted to the whitewashed beams and taught us how to do skin-the-cat and the monkey-hang, and sometimes led us in chinning contests (he was an agile farmer—we often rushed to the kitchen window after supper to watch as he stepped off the porch, kicked up his heels, and walked all the way to the barn on his hands). We'd see who could pitch the milk rag into the soapwater bucket from the farthest distance. He taught us to squirt milk straight from the cow's teat into the gaping mouth of a barn cat, which was entertaining for everyone involved except the cow. He told us Ole and Lena jokes, and we laughed whenever one of us got smacked in the face by a cow tail freshly swabbed through the festering gutter. Some nights we got to talking and the conversation ran all evening long, moving from cow to cow with breaks to dump the milk. The discussions were omnivorous, covering fishing, the price of corn, and once—I have no idea why, as Dad didn't talk sports, but the scene persists with absurd clarity—Green Bay Packers running back Terdell Middleton. On another night, Dad looked up from where he was kneeling beside the big Holstein and in a quiet voice advised me to beware the study of philosophy because I would wind up questioning everything including my own faith, and over time and in essence he would be proven correct, although a half-read secondhand copy of *Thus Spoke Zarathustra* does not a philosophy major make.

When the last milk was poured, Dad rinsed the milkers and hung them for the morning while I busted hay bales and kicked flakes the length of the mangers. After feeding the calves, I shook out fresh forkfuls of straw beneath each cow. Then I killed the lights and listened for a moment as the cows nosed through the hay and prepared to bed down.

We left the barn together then, pausing a moment to turn and check Orion's progress. He was above the barn now, just clearing the roofline, halfway through another all-night cosmic hurdle. Satisfied by the sight, we turned for the lights of the house.

Down here at our new place, I work in an office above the garage. While stumping the short distance across the yard to the house after writing late into the night, I often stop and study the silent structure, knowing my wife and daughter and the unknown unborn one are in there slumbering under the assumption that I have somehow been using the time to provide. Spinning a living from typing and talking and traveling is all well and good, but I can tell you the project does not bear up under scrutiny at 2:00 a.m. and ten below. Especially if you've just burned six hours and two pots of coffee tweaking a sentence fragment that holds together like cheese crumbles. Calvin Coolidge notwithstanding, sometimes persistence is just a batty cat slapping at a mirror.

I'm not trying to become the farmer my father was. I'm not even trying to become my father, although the parallels are lately multiplying. But I reconnoiter with his example constantly. Tonight I stand in the cold and study Orion for a long time. The first day I set foot on this place, I became one quarter-twist discombobulated and got it in my head that west was north. I know better now, but still encounter a fuzzy two-second delay when verifying my bearings. So it's good to see something familiar in the firmament. From Orion I pivot to locate the Big Dipper, which never leaves the sky. This too is a comfort. Tracing a line from the base of the dipper to the lip and beyond, I locate the North Star. Dropping straight down to the horizon, I shift my gaze a few degrees west, where forty miles north my father is asleep, his children gone about their business in the world.

CHAPTER 3

I am in the office working after supper when Anneliese calls. She is having contractions. "I think they're just Braxton-Hicks," she says, using the term coined for the nineteenth-century physician who left his name to false labor, "but they're coming pretty steadily." She is just over six months along, and I am immediately light in the chest. When I get to the house she is breathing through a contraction that has lasted over a minute. I go into full Evelyn Woods mode on the stack of birthing books I was supposed to have read months ago, fingertipping the indexes and speed-scanning everything I can on premature labor. Ten quiet minutes pass, then Anneliese says, "Here's another one." Another follows five minutes later. And yet another forty-five seconds later. Then another five-minute gap. Even as I'm reading pertinent sections aloud to Anneliese, I'm trying to convince myself that it is nothing, but I am not feeling brave at all. Then the cycles slowly subside. By bedtime nothing is happening. I am a worry champ, and pull the stethoscope from my emergency medical kit to double-check the baby's heartbeat. It's there, but I check it three more times before we are asleep.

I was raised in an obscure fundamentalist Christian sect. Our ministers (we called them "workers") divested themselves of all possessions and went forth two by two, spreading God's Word by means of gospel meetings held in village halls, bank basements, and American Legion posts. If you came as a stranger you would notice the quietness as the people gathered, removing their caps and hanging their coats without conversation before seating themselves in the neat rows of folding chairs the workers had set beforehand. Just inside the door one of the workers would pass you a copy of *Hymns Old and New* from an open briefcase at the back of the room. The loaner hymnals were the size of a thin *Reader's Digest* and bound in brown plastic, and they were lyrics only—no notes. Still, it wasn't tough to sing along. We took things slowly and kept a lid on it. If there was a piano available, some-one might play it, but soberly. One night a tough old farm lady named Florence took a seat on the bench before the aged upright piano in the Prairie Lake Town Hall. Florence wore orthopedic shoes and horn-rimmed bifocals and kept a hanky tucked in her bosom, but when she leaned into that first verse she laid a left-handed barrelhouse rumble beneath the praise such as we had never heard before. Our eyebrows shot up and we swung right along, delighting in the spirit of it. Afterward the older brother worker had a quiet word with her and sadly Florence cut the honky-tonk. Another time an itinerant evangelist showed up late for gospel meeting and crept into the back row with a tambou-rine, which is like showing up for a Gregorian chant armed with a pink kazoo. I stole glances over my shoulder every time I heard a muffled tinkle.

Ours was an invisible church—a church with no name, and a church that didn't believe in churches. We were the church. As the New Testament instructed. When it was time for Sunday morning meeting, we convened in private homes. To raise a structure and call it a church was the worldly way. A church made of hands was soon cluttered with altars and crucifixes, and was thereupon idolatrous. These false churches, they were not walking in Truth. They were whistling off to Hades. This was a shame, because I knew some real nice Lutherans. In conversation we spoke of each other as the Friends, and sometimes said we were in the Truth, but there was no letterhead anywhere with "The Truth" stamped across the top. When we said we had no name, we meant it sincerely. Yes, but it has to have a *name*, we would hear, again and again, as if we were playing a trick. Sometimes the outsiders called us names—the Two-by-Twos, the Dippers, the Black Stockings, the Damnation Army—but these were *outsiders*. Outsiders—as we were reminded at gospel meetings—were *worldly*. Not worldly as in "sophisticated." Worldly as in "set to sizzle."

Gospel meeting opened with hymns and a prayer. Then the younger worker preached for twenty minutes or so. After we sang another hymn, the elder worker preached the longer second half, and then after one more hymn and prayer we were done. The workers rarely brought the brimstone; rather, they generally spoke in a narrow range of tones somewhere between astringent history teacher and gentle physician. After preaching a town for a few weeks, they might "test" the meeting on the closing night of the run. During the last verse of the final hymn, anyone who hadn't done so previously was invited to stand and profess their faith in Christ. This was a serious step—in short, it meant you were officially joining up. You were now walking in the Truth.

I admit there are times while traveling in certain circles that I take some perverse joy in letting slip that I was raised in an "obscure fundamentalist Christian sect," because for some disinclined folks the phrase conjures a wild-eyed tribe of charismatic Bible-wingers hoarding automatic weapons and diesel fuel within a walled compound. When I reveal that I am no longer a member, there is the underlying inference that I escaped under cover of darkness and must forevermore avoid Utah. Sadly for the sake of cocktail talk, ours was a pretty low-key operation. No speaking in tongues, no Holy Rolling, and grape juice for communion. We kids went to public schools, our parents worked regular jobs, and at first glance the only thing you might notice was that our mothers wore dresses and stacked all their hair up in a bun. Mom did wear high-top construction boots with her maxi skirts, so *that* was a little offbeat.

In what can now be seen as some sweet irony, my mother was known in her youth to repeatedly state that she didn't care who she married as long he wasn't a farmer—in fact she once declined a marriage proposal from a barge worker after he told her he had saved nearly enough money to buy his dream farm. Mission accomplished, then, when in 1963 she wed my dad—a freshly minted chemical engineer with job prospects in Minneapolis. They met via the alphabet, which placed them proximal on the Eau Claire Memorial High School homeroom seating chart: Perry, Peterson. He first caught her eye with his assiduous study habits; every morning he'd take his seat, crack his books, and buckle down. She later found out that he was cramming for his first-hour Spanish class.

Mom was smart, proper, and devout, having professed her faith in the Truth at a young age. Dad was smart too, but he was a worldly boy, and although he was pleasant and quick to grin, he sometimes comported himself as a potty-mouthed hotshot. Short, small, and quick, he was a champion wrestler. He wore his hair buzzed close to his scalp. This accentuated his ears, which were small but curled outward in the manner of Frito Scoops. In testament to his skills as a grappler, the vulnerable ears were not cauliflowered.

My mother was shy to the point of pathology, but she did possess reserves. Once—unbeknownst to Mom—her socially active cousin placed her on the ballot for class secretary. Mortified when she learned of the conscription, my mother's immediate inclination was to decline, but then she decided she would just smile and say hello to everyone in the hall, and she was elected. My father claims this same proactive cousin attempted to set him and Mom up when they were freshmen, but Mom says she never knew of this.

All through high school then, my mother and father began the day together but never dated. On graduation night, they wound up at the same house, with a group of other students gathered for—as Mom once described it while rolling her eyes and shaking her head—"a *learned* discussion." When the colloquium concluded late in the evening, my father wanted to take a different girl home, but missed his chance when she left with another boy.

Dad drove Mom home instead. Somewhere on a dark road, he ran out of gas. Oldest trick since the invention of the internal combustion engine, really, except that he honest-to-goodness did run out of gas. They walked the last three miles. The county had recently graveled the roads and Mom ruined her heels. The young couple reached my mother's house at 3:00 a.m., and she woke her father to ask if she could borrow his car and a can of gas. Grandpa said sure, fine. He trusted her. By the time she got back home again, it was 5:00 a.m. At around that same time my father was waking his parents to explain where he'd been. They accused him of lying.

The same group of students gathered again the next night, but according to Mom the discussion wasn't as fun. That night Dad drove the other girl home.

The other girl didn't stick. Later that fall, when Mom was in her first year of nursing school and Dad was a freshman at the local state college, he asked her to homecoming.

In Mom's words, the date was "a great fiasco." She agreed to go to the football game, but as she was already a member of the Truth, which had strictures forbidding dancing, she refused to attend the dance. Furthermore, Dad had been drinking the night before, and was certain Mom could tell. She says he couldn't wait to get her home and off his hands. At the door, she invited him in for cocoa. I delight in the image of my dad blowing on that hot chocolate, his toes curled tight as a pipe clamp, sweating out the last of the previous evening's booze and just—I have to assume—dying for a real drink. He drank the cocoa and bolted.

One year later, they went on a second date. "This is getting *se-*

rious," said Grandma Peterson. And despite the slow start, it was. Within a year Mom was on her way to being smitten. But she was troubled: in the Truth marriage to outsiders was forbidden. And she felt strongly that shared faith was the most critical bond of marriage. Dad was a discontented Methodist, but when he asked to attend Sunday meeting with her, Mom told him no. She thought worldly people were only allowed at gospel meetings. That spring, she went to Mexico to visit a pen pal in Guadalajara. She had begun writing to him when she was twelve and he was fifteen. The boy was now a medical student, and engaged to be married. She found him pompous. But she liked his sisters, and enjoyed her time with his family.

Mom had begun praying for a good husband when she was very young, and in Mexico her prayers continued. But now she was praying for the strength to tell my father that she could no longer countenance dating him when she had no intention of marrying outside her faith. On the way home, she rehearsed her speech and redoubled her prayers.

Two surprises awaited her. The first was a letter from the Mexican medical student: he wrote that he had ditched his fiancée and intended to marry my mother. She could come to Guadalajara and be his wife, he said. She would also have to convert to Catholicism, but that was easily arranged.

The second surprise was more pleasant by a mile. With Mom away in Mexico, Dad went directly to her father and asked if he might come to Sunday morning meeting. Grandpa said sure. Moved by what he saw in the quiet gathering, Dad arranged to attend a gospel meeting. He was prepared in his heart: when the meeting was tested, he stood, committing himself to Christ, and, by default, to my mother. They were wed in September on my grandfather's front lawn against the backdrop of a trellis decorated by autumn leaves. Per Dad's request, the wedding cake was chocolate. In the portraits, Mom is a dark-haired beauty in a sheath dress, holding a spray of autumn mums. Dad looks like a spiffed-up little boy who won the pine box derby but could bolt the podium. Later it would be discovered that one of the attendants who signed as witness was underage, leading to the delightful possibility that despite their eminent respectability, my parents might officially qualify as shacked up. On those poignant occasions when someone hauls off and calls me a bastard, I peep furtively left and right, and then whisper, "*Entirely possible.*"

The newlyweds honeymooned in a rented cabin up north near Danbury, Wisconsin. Dad went fishing while Mom read books in the boat. Clearly there was dew on the rose—in the forty-two years since, I have never once seen my mother in a fishing boat. Upon returning to Eau Claire, the couple took up housekeeping in a small downtown apartment, and my father began a job search. Within a month Dad was hired by Archer Daniels Midland to study alternative uses for soybeans, and the young couple moved to St. Paul, Minnesota. Apparently Dad's was the perfect gig for a science geek—among other things, he experimented with reconstituting soybeans in the form of cheese curls and glue. Throughout my childhood there was always a large tin coal pail in the playroom. It was filled with laminated blocks, and Mom still keeps it in the living room for the grandchildren. I only recently learned that the blocks were manufactured during one of Dad's experiments—the laminations were held together by soybean glue. I am ill-informed as to the current state of regard for soybeans in the fixatives industry, but I can report that after four decades of grubby mitts and slobber, those blocks are holding fast.

During their engagement, my parents had applied to the Peace Corps. Commonly enough, they wanted to help other people. Specifically, Mom hoped to provide maternal and child health care in Central or South America. Shortly after President Kennedy was killed, they received their call. Dad left his three-month-old soybean-squeezing career behind, and they moved to Northern Illinois University to undergo their initial training. Two and a half months later, they traveled to Hawaii for a final session in preparation for being deployed.

Hawaii was beautiful. They were given time to travel, and in a sense it was a second honeymoon. Then came a surprise. "They told us women the shift to Hawaii would bollix up our menstrual cycles," Mom told me recently. "And sure enough, I missed one. Then, being an O.B. nurse, I noticed some other things. So I got a med tech friend to give me a pregnancy test."

And there I was.

And that was the end of the Peace Corps. They informed their group leader, and someone called Washington. Can't go if you're pregnant, said Washington. When Mom and Dad left home they said their good-byes, not expecting to see their friends and relatives for two years. They were back in four months.

Some nights in the farmhouse after the cows were milked and the dishes drained, Mom and Dad would gather us in the dark and show slides of their abbreviated Peace Corps stint. For a rural Wisconsin kid, the images from Hawaii were tantalizing—

volcanoes, gargantuan flowers, fields of sugarcane ablaze. Whereas all I remember of the Illinois photos was a handful of images showing tree limbs and power lines laden with ice. Mom said the ice storm was really something—it paralyzed De Kalb for days—but I'd witnessed the same thing in my own backyard and wasn't very impressed.

Then too, I came of age during a time when the finest thing your average frostbitten Midwesterner could imagine was a trip to Hawaii. How we envied those who ventured out pale from between the snow-banks only to return a week later looking like scorched beets in pine-apple shirts. "We were in *Hawaii*," they'd say, fishing a tin can from the depths of a Naugahyde *Aloha!* travel bag. "Have a macadamia nut!" Down at the café or the tavern or at family reunions, whenever conversation turned to wintertime vacation plans, Hawaii was sure to pop up. You always envied the ones who had made the trip. So over the years I worked up this bit: if someone asked me if *I* had ever been to Hawaii, I'd say not only have I *been* there, I was *conceived* there. I told the story many times, often in the presence of my mother. In all my life I have never heard my mother indulge in even the most innocent double entendre or off-color comment (the fact that she says "bollix" doesn't count, as I can assure you she is utterly oblivious of the fact that it is derivative of the mild English expletive *bollocks*, and when she reads this she will be *mortified*). Whenever I delivered the Hawaii punch line she would avert her gaze or pat her legs the way she does when she's uneasy. And then one day when I was well into my thirties, we were at a family get-together. Hawaii came up, and I reprised the bit yet another time. Mom motioned me into the hall.

"I know you enjoy telling that story," she said, patting her legs. "But it's not right."

Pat, pat, pat.

"I don't think it was Hawaii."

Pat, pat.

"I'm pretty sure it was during an ice storm in Illinois."

Back in Wisconsin, Mom and Dad bought a house and forty acres just outside the small town of Nekoosa, and Dad hired on at the Port Edwards paper mill. I was born at Riverview Hospital in Wisconsin Rapids at 1:42 a.m. on December 16, 1964. Before Mom returned from the hospital, Dad grabbed a swath of paper from the mill and made a sign that read WELCOME HOME MAMA AND MIKE. Right next to the word HOME he did a pen-and-ink sketch of our house—a log cabin that had been tacked over with off-brown faux-brick tarpaper. Dad taped the welcome sign to the old upright piano in the living room and placed a couple of baby gifts on the bench. Mom took a snapshot of the arrangement and glued the photo into my baby book. Just to the right of the piano is the rocking chair where Mom nursed me and Dad lullabied me to sleep. Two Bibles are visible on the music rest, stacked atop each other, the gilt pages lapped over the edge within easy reach of the rocker.

For a brief couple of months, I was a treasured only child. Then the other kids started coming, and for the duration of my childhood they *kept* coming. Just inside the front door of my parents' current house you will find a row of ten wooden lockers stretching fifteen feet from the welcome mat to the kitchen. Dad constructed the lockers himself and may have been in a rush, as the pencil marks are still visible through the varnish. Each locker had an integrated bench seat and separate spaces above and below the coat rack area for headgear, mittens, coats, and boots. Mom called the lockers "slots" and assigned us one each, using a grease pencil to inscribe our names above the coat hooks. Despite the nifty setup, the slots were forever overflowing with winter clothes and chore clothes and whatever we dumped after school, and Mom was continually admonishing us, "Clean up your slot!" which out of context sounds strangely personal. It was a losing battle. The porch nearly always looked like the back room of a Goodwill store under the inattentive management of compulsive ragpickers.

When you tell people you were raised in a large family, they come right back wanting a specific number, but we operated on a sliding scale. I have had a multitude of siblings; some born of the same womb, some adopted, some fostered, and some arrived in the nighttime absent formal affiliation of any sort. Some stayed for a weekend, others their entire lives. The last time my mother put a pencil to it, she calculated sixty or so children had come into her care. The one time we all sat for an official family portrait, in 1979, there were eight kids and two adults, so let's just say on average we were a family of ten. Or know that one night before supper in the early 1970s Dad put an extra leaf in the dinner table, and it never did come out. "Grab what you want the first time," he would say whenever we had guests at mealtime. "It ain't comin' around again." He replaced the chairs on one side of the table with a wooden bench upon which we sat shoulder to shoulder. Mom summoned us to supper by leaning out the porch door and rattling a cowbell, and we came from all corners.

I was five months old when Mom and Dad took in their first foster child, a five-week-old infant with microcephaly. Her name was Connie, and she was "pre-adoptive," meaning Mom and Dad

were to care for her until the county arranged permanent placement. Some time later the social worker told her Connie lived
just three months after leaving. Because Mom was a nurse, the
county also began sending her "special needs" children. The first
of these was a young boy named Larry. Larry was recovering
from rheumatic fever, and per doctor's orders was supposed to
remain confined to the couch. Today Larry would likely be diagnosed with some behavioral disorder or another, and his family
simply couldn't manage him. He came with holes in his clothes,
Mom says, and he was a handful, but full of fun.

Eventually as Larry regained his strength and was so allowed, he put me in a cardboard box and rolled me around the
house on my Playskool Walker Wagon. Then one day he pulled
me from the box, wrapped my fingers around the wagon handle,
and turned me loose. When I flopped, he picked me up and re-
launched me. Again and again we set out across the linoleum
tiles, Larry hovering as I stumped along to the rattle-jingle of the
balls and bells bouncing in the cylindrical cage of painted dowels
that spun between the wheels. Eventually he weaned me from
the Walker Wagon and turned me loose without props. One step,
a couple steps . . . again, every time I fell he would right me and
relaunch me until one day I just kept going. On average, I have
been toddling smoothly ever since.

I don't remember Larry, of course. In the photographs, he is
a gangly kid with horn-rimmed glasses and a big grin. Mom says
he would trap the cat under the sofa and then, employing the
cardboard tube from a roll of wrapping paper as a megaphone,
holler, "Come out, Kitty, with your hands up!" When he had re-
covered from the rheumatic fever, the county moved him back

home. Years later my parents heard he was injured in a bicycle or motorcycle accident, but they know nothing more than that. It's something, though, to study that black-and-white photograph of me in the box, him with holes in both pant knees, and think, somewhere out there—if he survived—is the boy who taught me to *walk*.

For a short time, it appeared as if my parents had settled in Nekoosa. Dad went to work at the mill in the morning, and cut firewood out back in the evenings. They were content at home, but Dad was dissatisfied in his work. Hired as a "research scientist," he spent most of his days at a desk with nothing to do but watch trains come and go. As he looked out his window he began to formulate the idea that he would be happier in the northwest part of the state. He had pleasant memories of visiting his uncle Robert, a farmer up near Spooner, and his family still went deer hunting in the area every November. When a job matching his qualifications became available at a small factory in Bloomer, Wisconsin, he took it. This was a little farther south than he and Mom were hoping, but when a farm fifteen miles to the north came for sale, they decided to take the plunge, paying $14,900 for the buildings and 160 acres—80 of it tillable, the rest swamp and trees.

And so it was that in June of 1966, the three of us put the noxious stacks of the Port Edwards mill in the rearview mirror of our '56 Chevy wagon and headed across state for a new life in the northwestern corner of Chippewa County, Wisconsin. To this day both Dad and Mom claim the motivation behind the move was to raise their children in the country—there was never any plan to farm. In fact, when I ask him about it now, Dad says, "I don't think I even realized I had that particular defective gene." He went to work at the factory in Bloomer, making $2.20 an hour. But within a year he got a half-dozen sheep, and not terribly long after that he drove over to the neighbors and came back with the milk cow, and despite all the best-laid plans, my mom became a farmer's wife.

One of the reasons we're having a baby is that Anneliese felt Amy should be allowed to grow up in the same house as a sibling. When she asked my opinion, I didn't really know what to say, having never known any other way. The first sibling I can recall was a girl named Eve. She had blond hair and cat's-eye glasses. I remember her pulling me in a wagon beneath the yard light beside a wild rosebush, although there is a black-and-white photograph of that moment in my baby book, and I wonder if I have animated it for memory's sake. Eve was yet another "pre-adoptive" child, and she stayed with us for a year before the county placed her permanently. My father says her last night on the farm was one of the worst of his life. She cried and screamed that she didn't want to go. I would only see her two more times—once a decade later when we were teenagers and she came to a nearby Bible camp, and once at her wedding reception. Both times it was wonderful to catch up, but so much time and life had passed that it was difficult to envision her as the sister I knew. In fact, while I can clearly remember her face from the days we played on the farm, I cannot summon it from either of the later two visits.

I don't know where Eve is now. Our last contact came fifteen years ago in Eau Claire, Wisconsin, where I had heard she was serving as a police officer. During a trip to the public library I got caught up in the stacks and returned to find I had overstayed my parking meter. When I read the signature on the citation, I recognized Eve's name, and thought fourteen bucks was a fair price for the fun of getting a parking ticket from a long-lost sister.

My brother Jud was mentally disabled (we simply used the term *retarded* in that age) as a result of complications at birth. He was the youngest of five boys orphaned when their mother died of cancer and their father subsequently shot himself in the basement, experiences that added a layer of psychological trouble to his preexisting problems. He was prone to fits of yelling and screaming, and occasionally ate his mittens. The day he arrived we celebrated with a rare stop at the A&W root beer stand in Chetek. We had a four-door Chevy Impala at the time and there were kids crammed front and back—Dad was at work, so Mom was driving. After the waitress fixed the tray to the window, Mom started passing out hot dogs, beginning with Jud, who was seated directly behind her. When all the dogs were in hand, Mom set to divvying up French fries. When she turned to hand Jud his portion, he was swallowing the last of his hot dog, napkin and all. Another time he devoured an entire bag of unpeeled oranges. For all his voracious eating, Jud was always thin as a rail, no doubt due to the fact that he never stopped moving. He wore out a series of wheelbarrows, and used to sit sidesaddle in a little red wagon and push himself round and round the driveway with the sides of his feet until his leather boots wore through. When given a book, he would page through it compulsively until it was shredded. Since he was so hard on books, every Christmas my grandmother wrapped the JC Penney catalog and gave it to him. It was his favorite present. He'd strip away the paper and start flipping through the pages, front to back. When he reached the end of the catalog, he'd flop the catalog over and start through again. My brother John and I shared a bedroom with Jud for a while, and we remember waking at 2:00 a.m. to the sound of the pages going *flip, flip, flip* in the dark. *Flip, flip, flip . . . FLOP. Flip, flip, flip . . . FLOP . . .* By the time next Christmas rolled around the catalog was in tatters.

In his teen years, Jud was tall and distinguished, with a shock of John Kennedy hair and a patrician jawline. When he was relaxed and his most obvious tics were suppressed, he projected an air of erudition. One evening a stranger drove into our driveway looking for directions to New Auburn. My brother Jed, then about ten years old, gave the man perfectly good directions. Just as he finished, Jud sidled up. "Go north. Two miles, take a right, then straight," said Jud, in a fractured recitation of Jed's directions. The result was utter nonsense—beginning with the fact that New Auburn lay to the south—but the way he rattled it off, it sounded believable.

Jed pointed up at Jud. "He's retarded."

"OK, little fella," said the stranger, chuckling and patting Jed's head. Then he climbed back in his truck, drove to the end of the driveway, and, exactly as Jud had instructed, turned north to nowhere.

During much of my childhood we double-, triple-, and occasionally quadruple-bunked. When my brother John and I slept in a converted closet at the top of the stairs, we could stand erect on only one side of the "room," as the other half was transected by the roofline. Per John's request (he now owns a dump truck and a sawmill and will deny this, but I can provide photos), Mom painted a butterfly on the slanted ceiling. It was an attempt to evoke spaciousness, but that just meant when you stood up, you smacked your head on a flat plaster butterfly.

With an eye to the expanding brood, Dad began to remodel the old three-bedroom farmhouse in the early 1970s and expects to finish the project any time now. There was always some wall being knocked out somewhere. Jed learned to climb ladders while still in diapers, and at one point when the ceiling was being reconstructed we amused ourselves by fishing for sandwiches through a hole cut in the upstairs floor. We'd set up an ice-fishing tip-up over the hole, lower the line, wait for the tug that released the flag, and then reel up a sandwich Baggie.

Eventually it became obvious that the house simply wasn't big enough, and Dad hired my uncle to help him build an addition that exactly doubled the size of the house. John and I were so excited at the prospect of having our own rooms that we would drag our sleeping bags through the second-floor window into the partially constructed addition and sleep on the subflooring with nothing but the naked stud walls separating us. Years later when I viewed reruns of *WKRP in Cincinnati* and saw Les Nessman delineating imaginary office walls by strapping tape to the floor, it reminded me of John and me sound asleep in the unheated addition, separated only by two-by-fours on sixteen-inch centers. When the addition was finished, the upstairs hallway was over forty feet long with nine doors.

The house was now officially bigger than the barn. We treated it as a combination amusement park and gymnasium. Mom had a no-running-inside rule, but beyond that she pretty much turned us loose. We tore apart the couch and used the cushions to build forts, and we used a cardboard refrigerator box to construct a submarine in the living room. Donning the flippers and snorkel masks Grandma Perry brought back from her vacation in Aruba, we'd belly-crawl out through the imaginary pressurized porthole and frog-kick across the linoleum, scanning the murky depths with the miniature flashlights that same grandma put in our Christmas stockings (Grandma Perry ignored the No Christmas rule and Mom and Dad let us). Mom kept the house stocked with art supplies, and often mixed up finger paint, which we swabbed across giant chunks of waxed paper torn from one of the rolls Dad got at surplus when he worked at the Port Edwards mill. I sat for hours at the play table looking at the bird feeder outside the picture window, drawing blue jays and evening grosbeaks, with Mom's copy of *Birds of North America* as my guide. We had a Visible Man (his halves held together with rubber bands) and we studied his visible liver, but like most kids, we were mostly interested in stripping out his skeleton, as it reminded us of Halloween. We used our Tupperware Build-O-Fun kit to cobble up Dr. Seuss–like vehicles, and passed snowbound winter mornings inhaling the scent of wood smoke from our Temp-O-Matic Woodburner set. When the wood-burning got tedious, we would "accidentally" jab the red-hot Wonder Pen into the Styrofoam packing and sniff the poisonous yellow smoke. We dumped out our Lincoln Logs, strewed our Tinkertoy Master Builder set from kitchen to porch, and used Mom's saucepans for army helmets.

On winter nights when it got dark early Mom let us turn out all the lights in the house and play hide-and-seek. I remember the giggly-scaredy feeling of trying to hold super-still when you were just about to be found, and the clatter of pots and pans as one of my siblings bailed out of the cupboard and made a beat-feet break for the in-free post.

When you grow up following a religion called the Truth, surrounded by the friends, and guided by the workers, some austerity is a given. At the top of the list, our church forbade the possession of televisions, which were condemned as the leaky end of Satan's sewer pipe. Prodigal though I am, I largely retain the sentiment, although honesty compels me to admit this has not stopped me from participating in the medium at both ends of said pipe, and compared with high-speed Internet, the boob tube has all the turpitude of worn-out View-Masters. Despite leaving the church in my twenties, I went for years without a television. When I got married there was backsliding, as my wife's dowry included a combination VCR/TV unit with rabbit ears that pull in four fuzzy channels. I justify its presence by citing PBS, but given half an hour, the snowy *Seinfeld* rerun triumphs every time. We go through fits of self-revulsion during which we banish the set to the closet and pull it out only to let Amy watch *The Magic School Bus* or a Lightnin' Hopkins documentary on DVD ("Mom!" she said when Anneliese walked in the room, "Lightnin' is *dead!*"), but then Anneliese has another couple of sleepless pregnant nights or I am feeling sorry for myself over some deadline or other, and whammo, it's *Scrubs* at midnight. By the time you read this the new digital format will be in play and our set will be worthless. We have sworn a solemn vow not to purchase a converter box and can use your prayers in this regard. The flesh is weak, particularly that mushy area directly behind the eyeballs. Church precepts were fuzzier regarding use of the radio, but Dad drew a firm line against it. One of our Volkswagen buses came equipped with an AM radio and I recall sneaking out for a listen, but in the process of trying to improve reception I reached beneath the dash and wiggled some wires, whereupon there was a blue flash, a whiff of scorched electronics, and the radio was forever rendered mute.

Perhaps allowing the devil a toenail in the doorjamb, Mom kept a phonograph in the house, and with her permission we were allowed to play it. The cabinet contained albums by Pete Seeger and the gospel singer Evie, a *Reader's Digest Presents 50 Beloved Songs of Faith* collection, and five or six mariachi albums from her time in Mexico, which would explain why someone passing through rural Chippewa County in the early 1970s might have heard the sounds of a *guitarra* and our preadolescent Scandihoovian voices yodeling, *"Ai-yi-yi-yiii!"* These were the only words we knew, although my brother Jud, whose mental disabilities were leavened with certain savantisms, including the ability to memorize entire record albums after just one or two listens, sang along in phonetically serviceable Spanish. When we put on Stan and Doug albums, Jud switched effortlessly to a Scandinavian accent and recited the goofball tunes word for word. We had a smattering of 45s—I remember distinctly "Heartbreak Hotel" by Elvis and "Green, Green," by the New Christy Minstrels. Many of these were leftovers from the day my teenage Aunt Sal brought her shotgun to the farm and practiced skeet shooting, using a stack of her "old" 45s in place of clay pigeons. I know there was more Elvis and plenty of Beatles in that stack, and shudder to think that somewhere in the subsoil out behind Dad's barn are the irretrievable shards of an eBay bonanza sufficient to finance Amy's pending orthodontia.

We also had six albums by Herb Alpert & the Tijuana Brass. They were easily my favorites. My father still had the trumpet he played in the Eau Claire Memorial High School band; I would pull it from the case and blow air-trumpet in sync with "A Taste of Honey" and "Bittersweet Samba." I have those albums now, and sometimes I load them on the old console stereo I keep in my office just for the pleasurable rush of memory the vinyl gives— "Green Peppers" puts me back in the old farmhouse, the brass notes echoing from the cool plaster walls, as the barnyard lies still beneath the noonday sun. I note that all three of the songs I cite are from the album *Whipped Cream & Other Delights*, which featured on its cover a lady wearing nothing but confection. The album was released in 1965, and millions of young boys have yet to recover. It was quite a deal to be riffling past the original cast *Sesame Street Book & Record* album and Mitch Miller's *Sing Along with Mitch* only to come face-to-face with such dairy-based profundity. If you held the cardboard sleeve at an angle you could make out just the hint of the curve of one of her mysterious naughty bits, and the naked implications blew my youthful fuse. I'm surprised Mom didn't cover the woman in duct tape, because when she discovered that the version of "Bill Grogan's Goat" included on an anthology of train songs featured a mild expletive, she took a stick pin and cut a groove from the beginning of the song to the end so when the needle hit that track, it skidded right past with a scratchy rumble. If you were to confront my mother and accuse her of censorship, she would reply, "Exactly."

Still: *Whipped Cream & Other Delights.* When I learned some thirty years later that the whipped cream was actually shaving cream, it did absolutely nothing to cool my jets.

Dad didn't care for the music, and when we heard the porch door open we turned it off, but he did sit at the upright piano sometimes after milking to plunk out hymns and then send us up the stairs with a remarkably groovy interpretation of "On Top of Old Smoky." For several years I rode my bike two miles for piano lessons over on Highway F with Mrs. North. My parents hoped I might one day be good enough to play hymns at gospel meeting, but I peaked with a workmanlike version of "Let There Be Peace on Earth" at the elementary Christmas concert, and when I discovered football the piano lessons petered out. To this day, however, thanks to Mrs. North I can read the treble clef just well enough to help Amy with *her* piano lessons.

My mother taught me to read when I was four years old. Mom is a compulsive reader. She reads for pleasure, she reads to edify herself, but more often than not, she reads because she can't help it. I understand. The minute I find myself sitting still, I start rummaging around for printed material. Pretty much anything will do—a book or magazine, sure. But also cereal boxes, the weekly shopper, the underside of the Kleenex box, or the back of the toothpaste tube. (I can recite by heart: "Crest has been shown to be an effective decay preventive dentifrice that can be of significant value when used in a conscientiously applied program of oral hygiene and regular professional care.")

As a toddler, whenever I saw Mom reading, I bugged her to read to me. And she did. Every day. One day as I pestered her with my copy of *Winnie-the-Pooh* while she was settled with a book of her own, Mom set down a rule: She would read one chapter of *Winnie-the-Pooh* aloud (this was the original text-heavy version, not the picture-book version), but then I had to sit there quietly holding my book while Mom read a chapter of her book to herself. It worked, and became standard procedure. It took me years to recognize the power of this gift: Mom taught me to love the idea of sitting quietly with a book long before I could make out the words on the page.

In time I began to recognize letters and make attempts at small words, so Mom sent away to a Chicago newspaper for a phonics book. When it arrived, she started at the beginning and worked through page by page (sample lesson for *C* and *K*: "This cat has a bone caught in his throat and he is trying to cough it up, so he says K-K-K as in Cat and Kitty"). Soon I could read on my own, although not infallibly. Dad tells the story of me pointing at the tailgate of the neighbor's pickup and saying, "F-O-R-D . . . *TRUCK!*"

During that same tumultuous third-grade stretch when I was getting religion with the help of Hazel Felleman's poetry collection, Mom was sorting through a box of secondhand clothing when a copy of *All Quiet on the Western Front* tumbled out. I took it to the porch, settled into a chair, and dove in. I'd love to say reading Erich Maria Remarque at the age of nine stood as evidence of a precocious literary bent, but I'm afraid it had more to do with a young boy's fascination for all things war. Whenever Mom took us on our regular trips to the Chetek Public Library, my brother John and I headed straight for the aviation section, raiding the stacks for everything we could get our hands on about the Red Baron, the French-American hero Raoul Lufbery, and our Ace of Aces, Eddie Rickenbacker of the Hat-in-the-Ring squadron. We put together glue-splotched and imprecisely decaled plastic models of Sopwith Camel biplanes and Fokker triplanes and strung them from our bedroom ceilings using black thread from Mom's sewing box. We entertained visions of ourselves running across the green grass of a sun-soaked British airfield, prepared to buzz into the fluffy white clouds where war seemed to be a romantic romp in the clouds, with a tip of the hat to the hail-fellow-well-met set to shoot you down.

I was drawing a lot of ornate battle scenes at the time, often at the elbow of another recently acquired pal of mine, Eric Jakobs. Yin to Hardy Biesterveld's yang, Eric was the well-behaved son of the local Lutheran pastor. He arrived partway through third grade and moved away not long after when his father was called to another parish, but for a stretch there we were best friends to the point that we created our own hieroglyphic secret code, the key to which we sketched out and buried in a tuna can near the culvert just up the road one evening when Eric was visiting. I hid it well, because when I returned on a decoding mission a week later, I couldn't find it. The culvert has long since been replaced, so who knows where the can wound up. Perhaps one day it will surface to baffle interstellar archaeologists.

Eric was a talented draftsman. In fact, his arrival knocked me from my position in the class as "best draw'er." I clearly remember looking at his stuff and feeling a seeping twinge of envy, but also thinking, Wow, he's better than me. Our works were sweeping panoramics in which the skies were clogged with ball-turreted B-29s, Luftwaffe dive-bombers (the Stuka was a favorite—we loved the aggressive geometry of the inverted gull wings, plus we thought it funny that a warplane might be branded a "Junkers"), P-51 Mustangs (consistently sporting shark teeth), and P-38 Lightnings. We scrambled a lot of those P-38s strictly because we fancied the exotic twin-booms look. On the ground, Panzers squared off with Shermans, and the guys in green sniped, machine-gunned, and lobbed grenades at guys in gray or black. We perfected our rendering of the German helmet with its visor and dropped rim (we secretly found it sharper-looking than the standard American GI soup pot) and carefully labeled every piece of enemy equipment with a swastika—an emblem we memorized with creepy assiduousness so as not to have the arms bent in the wrong direction. Every visible muzzle—on the planes, on the tanks, at the end of each rifle—spouted jagged flame. On an optimistic note, if a plane was smoking toward the earth, its pilot would be visible in the sky, parachuting safely to the ground. Perhaps an accidental archivist will one day prove me wrong, but as I recall there were few if any dead soldiers, and none of them wore green. War poured from our colored pencils not as hell, but as a circus plus fireworks where at worst the good guys suffered nonterminal flesh wounds. It was in this mind-set that I first read *All Quiet on the Western Front*. I still have the actual book. It's a 1930 hardcover edition and the gray fabric is splotched with some unidentifiable spill. From the first page, I cherished the characters. I loved the rough Tjaden and his lice-popping oven. I hated Himmelstoss. I couldn't wait to see what the witty scavenger, Kat, scrounged next. But I especially cared for the narrator, Paul Baumer. He seemed calm, thoughtful, and strong. I read him as just another steady Louis L'Amour cowboy. Then I got to chapter 9, and Paul stabbed an enemy soldier to death. He said the soldier was French. This did not compute. I backed up and reread the passage. From reading all those air ace books I knew the French were on our side. And were thus the good guys. But I had been operating under the assumption that the narrator was the good guy. He seemed like the good guy. He *was* a good guy. I puzzled over the section, rereading it several times to see if I had missed something in the chronology. And then it slowly dawned on me. Paul Baumer was one of the *bad* guys.

From an adult standpoint, my misread seems ludicrous. After all, three paragraphs into the book Baumer speaks of the "English heavies" hitting his company with high explosives; there are all the German names and surnames; and there are battle scenes with the French earlier in the book. I remember some of this niggling me at the time, but I was reading full speed ahead and pushed it aside, figuring I had missed some twist of history. But when I got to the scene in the shell hole, I could no longer get around it: Paul Baumer was a *German* soldier. He had killed one of the good guys. What did that make Baumer?

I don't keep a chart or anything, but to the best of my recollection I have read *All Quiet on the Western Front* seven times. As a boy raised on Bible passages, I can't say that it is the most important book in my life. But the impact of Paul Baumer's story was profound, if subtle. When I opened the book, I possessed the vocabulary necessary to read the book, but until that section in the shell hole I lacked the insight required to see it as anything but a good yarn. I began the book a third-grader believing all the good guys played for the right team. Now I was faced with the knowledge that a good guy might wind up on the wrong team.

I'm glad I had a friend like Eric Jakobs. He taught me a nice lesson in humility. He was a better draw'er than me. Period. He taught me what it's like to realize you aren't the best at something, and no amount of positive thinking or self-esteem building will change that fact, and you better figure out a way to live in light of that fact because other instances are pending.

A woman recommended by our midwife has come to the house to give us birthing instructions. It is a cold day, but the sun is shining warmly through the window and spotlighting the carpet of the living room floor, where we are pretending to have a baby. The instructor has been very thorough, and it is neat to receive instruction right here in our home. At one point she puts Anneliese on all fours in a stance intended to relieve lower back pain during labor. Then she rotates me around back in a massage position, and Anneliese and I get the giggles because, without putting too fine a point on it, the maneuvering reminds us of how we wound up in this situation in the first place. When the instructor leaves, I fear she may be upset with us over our lack of seriousness, but what she may not realize is that this hour on the carpet has been the best date Anneliese and I have had for months. It has been too long since we had a conspiratorial giggle. Last month I bought a card with a line drawing of a beautiful lady in a red backless gown. Today I took colored pencils and

put a round red belly on the lady, then two valentine hearts—
one hovering above the lady's chest and one tinier one above the
curve of her belly.

When we married, I was a bachelor of some thirty-nine years.
Anneliese was a single mom raising a three-year-old while teaching Spanish at the university. We met in a public library when I
was seated at a table selling books. I carry an abiding image of
Amy's pale blue eyes looking up at me and her mother's matching pair just above. For our first official date we met in a coffee
shop, talked forever, and then took a long walk that is currently
approaching its fifth year. While I took some ribbing about the
evaporation of my singletude and gave up my New Auburn address, it is Anneliese who is bearing the brunt of change: leaving
her teaching position, carrying the baby, homeschooling Amy,
and tending our new place the many days I am away or seques-
tered in the office. I love my wife for her willingness to take these
leaps, her strengths where I am weak, the way when she smiles it
is utterly without reserve, and yes, her clear blue eyes, as startling
this morning as when I saw them in the library that first day.

She has been caught off guard by the difficulty of this pregnancy. When she was carrying Amy she spent a month hiking in
Central America—at one point climbing a volcano. She experienced none of the persistent weariness, or the spates of contractions that come and go. Her belly is big now, and she walks with
her shoulders back to counter the weight. I watch her sometimes
when she doesn't know, and just like when I sit down to write her
a card, the close study precipitates a sense of pleasant wonder
that I have a wife and this is her. Last night we went out to eat
with friends, and it was good to see Anneliese laughing in conver-

sation. While we were waiting for the food to arrive, Anneliese and I held hands beneath the table, and at one point she gave my wedding ring a little wiggle just like when we were first married and couldn't quite believe it. When we left the restaurant we held hands again and she leaned her head against my shoulder as we walked to the car and I opened the door for her like any good boy on a first date would.

The big farmhouse in Chippewa County is mostly empty now. Mom and Dad still provide respite care for profoundly disabled children, and they have full-time responsibility for Tagg, a boy who was two months old when his drunken uncle shook him violently. Tagg's injuries were devastating—he cannot dress, feed, or care for himself, he cannot speak, and he is prone to outbursts of hitting and biting. The county asked my parents to provide temporary care until the court case was resolved—eleven years later he remains in their home.

The last kids to leave the old-fashioned way—by graduating from high school—were my sisters Kathleen and Migena. Kathleen joined the family when she was three months old (I remember her foster parents handing her through the door of our Volkswagen bus in a basket). Migena's route was far more circuitous. When her brother Donard arrived in New Auburn as part of a foreign exchange program, well-meaning citizens bunked him with another Albanian exchange student, not realizing they were from opposing factions in Albania's rapidly escalating civil war. To preserve the peace, Donard moved in with

Mom and Dad, after which they discovered his paperwork had been forged as a means of moving him safely out of the country. One night after Don had settled into school and become a familiar face at the table, Mom and Dad's phone rang. It was Migena, in tears and calling from Michigan. She had made it to the United States with another student exchange program, only to discover upon arrival that the Michigan program would not provide her with graduation credits, and thus no opportunity to continue her education in the United States. Mom and Dad drove to Michigan and brought her home.

Today Don and Migena have both gone to college and found good jobs. They have endured tortuous paper chases in order to maintain their legal status, with my parents filling out form after form and vouching for them with government officials as necessary. They show up at birthdays and holidays, and once even managed to get some of their relatives into Wisconsin to go deer hunting with us. I am proud to call them brother and sister. "How many kids are in your family?" people still ask, and now you know why I never have a number. We're all spread out now, over geography and vocation. I have a sister Lee in Montana. My sister Suzy served in the army and is now raising a son and taking college classes. Jud lives in a group home and I haven't seen him in years. Somewhere out there I hope Eve is well and Larry is walking strong. Back home, Mom and Dad are living the empty-nest syndrome writ large. Dad doesn't seem to mind. Shortly after Kathleen and Migena graduated, I visited the farm. "I got up the other morning and a miracle happened," said Dad. "For the first time in forty years, I was first in line at the toaster!"

Once I asked Mom what drove her and Dad to start taking in children all those years ago. I was expecting some philosophical and possibly faith-based answer. Not so. "When we were still dating, your dad told me he wanted sixteen kids," she said. I chuckled. "No, really!" she said. "I said, No *sir*! *Six*, maybe, but not *sixteen*! So we decided we wanted to have some of our own kids, adopt some, and take some in through foster care. It was just always our plan." Her own parents had begun taking in foster children when their youngest daughter—my aunt Annie—said she wanted a younger sister. Mom says she was reading the newspaper at the time and found an ad seeking local foster parents. She showed it to her mother—my grandma Peterson—and shortly they took in their first foster child. Grandma kept a photo album of each child she and Grandpa fostered, and when she died there were twenty-eight children in the book.

Even with the baby yet to be born, Anneliese has brought up the subject of adoption and foster care. I once heard a man say that when a woman asks, "Honey, do you think we should have another baby?" he might as well start setting up the crib, but I'm not sure where this will go, or if. It has not been one long gauzy shot for my folks. You cannot take in that legion of children over the years and find joy with every one. Many arrived with their own history of troubles. There were the runaways. There were the children returned to abusive homes through error and faulty oversight. On the upside, I think my siblings would mostly agree that our full house seasoned us to accept the unusual as usual. We were often perplexed by people who were uncomfortable or even fearful in the presence of an obviously mentally disabled individual, since we had learned to assume that if someone was barking at their macaroni, they *always* barked at their macaroni. Then again I also learned hard lessons about my own character, angrily sticking up for one of my sisters when a kid in the library teased her about her disfigured eye one day, then mocking and tripping her cruelly the next. In the ever-changing cast of tykes carrying damage—congenital, traumatic, physical, emotional—we came to see that even in the midst of our own warm childhood, all was not well everywhere.

There is every reason for me to emulate my parents, but I am hesitant because after watching them for my lifetime I know exactly what the workload entails, and I'm not sure I'm up to it even on a small scale. I am keenly aware of what it cost to provide us that rich life. My sister Rya arrived on a day when we were making lumber. I remember walking in for lunch with the rest of the sawmill crew, standing on the porch steps sweeping sawdust off my jeans, then walking into the kitchen to find a tiny blue-tinged baby asleep in a bassinet beside the table. Rya had Down syndrome, and the blue tinge was caused by a congenital heart defect common in Down syndrome children. In Rya's case, the cardiac issues were further complicated by a lung disorder.

For the next five years, as Rya underwent a series of surgeries and hospitalizations, our house took on the trappings of a pediatrics ward. Green oxygen bottles lined the porch. A lazy Susan on the counter was covered with medication bottles and there was digitalis in the refrigerator. Mom was forever coming from and going to doctor appointments and blood draws. Some of Rya's more serious surgeries were done in the University of Wisconsin teaching hospital in Madison. I accompanied Mom on many of these trips while Dad stayed back to milk the cows and care for the rest of the crew. Once when Rya took a precipitous turn for the worse I remember holding oxygen on her as Mom drove through a blizzard to the emergency room. In addition to changing Rya's diapers, we kids learned how to change her dressings and inspect the sutures for signs of infection. When my sister Kathleen was a toddler in footie pajamas she would tip Rya across her lap, cup her hands, and clap up and down Rya's upper back to loosen the congestion, just as she had seen Mom and Dad do.

Rya learned to speak only a modest few words, and even those were difficult for anyone outside the family to understand, but Mom expanded her ability to communicate by teaching her some sign language. She loved to clown—after going to a high school track meet, she would stand by the kitchen sink, raise one finger to the sky, go "bang," and then do a high-speed toddle across the kitchen floor, laughing all the way. She entertained us as much as her heart and lungs would allow.

In the end, it was her lungs that failed her. She needed more and more oxygen. When she could no longer get enough air to sleep lying down, Dad built her an inclined bench from boards he had sawn himself and she took to sleeping sitting up, resting her head on her thin arms, her favorite doll by her side with its own oxygen mask. In the background the oxygen humidifier bubbled with a sound that suggested a brook in a meadow. The doctors said there was nothing more to be done, and finally they were right. Soon she required round-the-clock care. Dad sold the cows—the family's only source of regular income—so he could split shifts with Mom. The night before she died, we were all in the living room around the couch that had become her bed. She was smiling widely and enjoying our company. At one point I went to the kitchen, and when I looked back in the living room I saw that she had removed her oxygen mask and clambered down from the couch. She was making the rounds of the room, spending a moment with each sibling. Eventually she toddled into the kitchen and found me. We went back in the living room then, and that is what I remember, our whole family gathered around and Rya with her mask back on, her breath a pulse of fog against the transparent green plastic, and in the morning she was gone.

On August 8, 1977, I rose from a metal folding chair in the basement of the Moose Hall in Barron, Wisconsin, during the closing verse of "Close Thy Heart No More," and committed my life to Christ. Sometimes when folks professed they rose with joyful weeping. Other times their faces would be twisted in some combination of relief and holy fear like Sam at the end of Robert Duvall's *The Apostle*. But although my heart was beating high I was composed, because I had been thinking about this for a long time, and I was ready. The conversion had been under way since the day I read "The Hell-Bound Train" and gave up cussing, and was herded to its conclusion by the Four Horsemen of the Apocalypse during a stretch when I got to reading The Revelation of St. John the Divine in bed alone at night. To paraphrase Townes Van Zandt on the blues, after Revelation, everything else is just *zippity-doo-dah*.

I believed, and believed fully. And when—many years later—my belief turned to doubt, I left the church the same way I came in—quietly, over time. I have no cataclysmic story to tell, no single precipitating crisis. I can summon a little residual crankiness over the usual anecdotal complaints—workers running folks off over matters of hemline and haircut, pious elders with televisions hidden in the armoire—but I would never cut it as a bitter heretic. By and large, the people I worshipped with were a humble, tolerant bunch, content to pursue quiet example over thunderous harangue. So much so, in fact, that when in my wandering I hear someone snarking on fundamentalist Christians, my first thought is, Hey—those are my people you're talking about.

When you drift as I have, the Friends call it "losing out." Lately I wonder if I was out before I was in, if the voice of Paul Baumer put an existential whisper in my ear, casting shadows between black and white. I wonder too what sort of self-pitying train wreck I might have become had I not been raised by two people whose daily actions transcended my dogmatic quibbles and still do. Sometimes Mom apologizes to us kids, saying she and Dad took on too much, and that we suffered as a result. She says this, and I think of all the books, and prayers before meals at the big table, and the parade we made trooping up the stairs to bed while Dad played "On Top of Old Smoky" one more time, and how cozy it was with ten of us crammed in the Volkswagen after gospel meeting on a winter night. Or how after twenty years of opening my emergency medical kit, the first thing I think of when I see that green bottle is Rya on her last night bravely beaming.

Anneliese and Amy have bundled up and gone cross-country skiing out the ridge. Two days ago we had a blizzard that laid down a thick batting of snow. The spruce tree limbs remain bent beneath daubs of white, and the wind has pushed a four-foot drift around the garage and right up to my office door. While the flakes were still dropping, Amy celebrated with repeated swan dives from the top step of the office stairwell, planting her face in the peak of the drift and chewing snow.

I can see them now from my office window, gliding back to the yard in the fading light. Amy is leading in her blue snowsuit and goggles and Anneliese coming up behind, her current state betrayed by just a hint of top-heaviness beneath all the bundling. Moments like this, when I see the two of them together at a distance, I often think of the three years of history they have on me. It's not unsettling; it's just one of those hiccups in perspective that can leave me momentarily disoriented. I shut my computer down and head for the house. We've planned an evening together, watching *Willie Wonka and the Chocolate Factory* on Satan's glowing box. In the kitchen Amy is apple-cheeked and giddy. "Watching" is a rare treat, and she bounces back and forth between the refrigerator and the counter, helping me put together a tray of cheese and vegetables while Anneliese pops corn.

Upstairs we settle in on a mattress, our backs propped against a stack of pillows. Amy snuggles in between us, trilling with happiness. After three years of being a visitor in this house, I'm still getting used to the idea that we live here now. I think of my parents in that '56 Chevy, leaving Nekoosa. As the movie begins and Amy turns her attention to the screen, I reach an arm around Anneliese and pull her closer, squeezing Amy between us.

CHAPTER 4

Winter is on the fizzle, and Mister Big Shot is looking for love.

Mister Big Shot is a cock pheasant. He has been appearing at the edge of our yard nearly every morning for several weeks now, and he is plainly addled by love. He sports a glorious set of head feathers: blood-splash eye patch, bottle-green Batman cowl, a pristine white collar. The colors startle the eye, bright in the brown weeds like scraps of birthday balloon. Sadly, once he follows his beak out of the weeds, Mister Big Shot reveals the limits of his machismo, because somewhere along the line the poor bird lost his tail feathers. You have to figure the bobtail is a drawback on the dating scene. Like a bachelor with a bald spot, he must find ways to compensate. And so he inflates his chest, struts the perimeter of the yard, and crows blusterously.

Thus we call him Mister Big Shot.

The first time I saw him, I was stepping out the front door after breakfast. He had emerged from the row of spruce trees beside the pole barn. I froze and whispered over my shoulder to Amy, *"Come here, look, look!"* I cautioned her to move stealthily, not wanting to scare him off.

Turns out we couldn't scare him away short of a shotgun. The relationship has gone from breathtaking Animal Planet moment to *There's that knuckleheaded pheasant again.* For all my would-be woodsy knowledge, it took me a few sightings before I caught on: *Wait a minute . . . isn't he supposed to have tail feathers?* We didn't have a lot of pheasants around when I was growing up, so I tracked down some pheasant photographs on the Web to check myself. Sure enough. Most male pheasants have grand plumage sprouting out their hinders—sweeping quills of the sort you might use to sign ceremonial parchment, or to accessorize your Robin Hood cap. I wrote to a wildlife biologist and asked what might have gone wrong. He told me the feathers could have been snatched by a predator in a near-miss. Also, he said, sometimes the tail freezes to the ground during cold snaps, and when the pheasant takes flight, some of the feathers remain fixed. I picture the pheasant windmilling like mad, getting zero lift, then—*puh-luck!*—he blasts wide-eyed skyward. The biologist also said if the pheasant was pen-raised, it might have broken its tail feathers while tussling with other pheasants. Mister Big Shot does seem a little too tame for his own good (we can get pretty close before he bolts), so perhaps he was raised by humans. On the other hand, the biologist said only 10 percent of released birds survive the winter, so in that case Mister Big Shot would have earned the right to strut.

In the process of our speculation about the missing tail feathers, I tell Amy the legend of how the bear lost his tail: Bear's friend Fox convinces him he can catch a fish by dangling his tail through a hole in the ice. Bear sits there all night long. In the morning he feels a nibble, but when he leaps up, his tail—which has frozen in the ice—is pulled off. A gruesome story, as many fables are. Amy draws a connection to the plight of Mister Big Shot, and we discuss whether or not he might have been ice fishing. Then Amy asks me to pretend I am Mister Big Shot at the moment he did the power-molt. I flap my arms, wince with feigned effort, then holler *"Yee-owch!"* and look behind me in dismay and wonder. Amy laughs and asks me to do it again. But then she goes sober on me. "Will he ever get his feathers back?" I tell her the biologist said they would grow back in August. Until then, we will know the bird when we see him.

Long before my father had cows, he was a shepherd.

One of the Friends by Nekoosa had sheep, and Dad says that's where he got the bug. When he moved to the farm in 1966, he began to gather the flock. He got four ewes from his brother-in-law over by Hillsdale, and bought another four from a local man named Earl. Earl wanted thirty-six dollars. Dad wrote a check. Earl looked at the check and then he looked at Dad, and then Earl said, "This better be good." Dad reminded me recently that this used to be sheep country. "Lots of people used to have sheep," he said. "The Mareses, Norths, Skaws—they all had sheep." He's right. I tend to recall my farm childhood through the frame of my youth—when it was dominated by classic family dairying. I had forgotten the early days when farming was still a patchwork endeavor holding the line against the narrower specialization to come. I start working my way around the repurposed or vanished farmsteads all around the township, and sure enough, I do remember these people having sheep. "You sold your milk all year round, in the winter you logged, and in the fall you sold your lambs," says Dad. In other words, you kept a diversified portfolio.

Ask my father what he "does," and nine times out of ten he will reply, "I'm just a dumb sheep farmer." But listen to a dumb sheep farmer for long, and you'll realize the self-deprecation (rooted in the relative unlikelihood that sheep will put you on the fast track to the Forbes 500) does a poor job of masking some underlying affection for husbandry. My dad, a man not given to pet names, often refers to his "woollies," and once when someone suggested that sheep weren't too bright, Dad responded with a question and answered it himself: "Y'know what you get when you inflate a sheep, paint it black and white, add two faucets, and remove its brain?"

He waits a beat. "A cow."

From my largely oblivious childhood perspective, Dad's sheep were a sideline, whereas cows required our daily attention. When you weren't working with the cows you were working on chores predicated on cows. The sheep were just always there. Gray lumps in the distant pasture, only remembered a couple times a year when we rounded them up for worming, or shearing, or when we cut the buck from the flock.

In the early part of the year, however, the sheep begin to wedge their way back into the schedule until they dominate. In February Dad sets up the pens and feeders and gathers the flock to be shorn, after which—having lost their winter coats—they take up permanent residence in the lambing shed until spring and the grass return.

Lambing season amounts to a month of insufficient catnaps. You tromp to the sheep shed every couple of hours, around the clock, four weeks straight, until the last ewe delivers. The thing you're looking for is a sheep showing signs of imminent birth. She may be pawing the straw, circling, or simply looking distracted. A ewe experiencing the early twinges of labor will sequester herself off along a wall or in a corner. At the onset of a contraction an otherwise placid animal will extend her neck, raise her head, roll back her upper lip, and wrinkle her nose. A laboring ewe will grunt softly, as if she is being nudged in the belly (I hear a chorus of female voices: *As she is, Einstein!*). Another means of early detection: Put out fresh hay. As her compatriots rush the feeders like woolly pigs, watch for the ewe who remains apart— she's next.

Midwifery-wise, your basic job is to stay out of the way. Observe from a quiet remove and let nature take its course. Recede. Wait.

Anneliese is having a rough time of it. She just can't find her way to sleep. The weariness shows in her eyes every morning, and the best I can come up with is a hopeful "How did you sleep?" This only forces her to confirm that she didn't sleep well, while I stand there like a fence post. Tonight when I come in the house Anneliese and Amy are watching a video, which is a sign to me that Anneliese is worn out. We put Amy to bed. She closes her eyes and wriggles happily when I tuck the quilt beneath her chin. She is getting so long, so tall. I follow Anneliese to bed, where I rub her neck and lower back. Then I massage the area over her uterine ligament on the left side, and when my hand crosses over, I feel the little being within hiccup. It makes me chuckle aloud, but it's also a jolting reminder of how while I meander around thinking of the baby in largely exterior terms, Anneliese lives daily with this life nestled inside her. I kiss her good night and turn to my side of the bed. Our midwife has lately recommended Anneliese drink valerian tea, much revered by the herbal set for its soporific properties. So far it hasn't helped, but there is a mug of it cooling on the nightstand. This I know: valerian tea smells like bad feet and overheated muskrat.

Lying in the dark, trying to ignore the valerian stench, I wonder how I'll do when the baby arrives. Whenever people find out we hope to deliver at home, someone invariably brings up the fact that I am a registered nurse and have worked as an emergency medical responder for twenty years. "You'll be fine!" they say. Then I tell them that in all those years, I have never seen a baby born, let alone delivered one. The only babies I've ever caught emerged from a pair of truncated plastic hips strapped to a library table during our biannual emergency responder testing. Those babies are plastic, and their umbilical cords are attached with a metal snap. Anneliese has stacks of beautifully written home birth books she wants me to read, but so far I've spent most of my time reviewing the very straightforward illustrations included in the obstetrics chapter of Nancy Caroline's *Emergency Care in the Streets.*

I spent four years in a fine nursing school, but my maternity rotation was a bust. Every time I went to the hospital, I got all prepped and ready, but never once was a baby born on my shift.

The only significant experience I recall was when my instructor asked a woman who had already given birth if she would agree to allow a student nurse to perform her "five-point checks." Five-point checks are an examination performed on the mother in the hours following childbirth to detect any abnormalities or problems. Three of the five points qualify as personal and specific, and a uterus massage is included.

"Hi," I said, walking into the room. "I'm here to do your five-point checks." The woman's eyes widened.

"Who are you?"

"I'm your student nurse."

"Oh. My. God."

I retreated half a step. "If you'd rather . . ."

"I said I didn't mind a student nurse, but I . . ." She trailed off. Then she took a deep breath and rolled her eyes. "Oh, what the hell," she said, hiking up her gown. "It's my third kid. Get it over with."

These mornings as soon as breakfast is finished and before I head up to the office, Amy and I collect maple sap. It's a small operation—just six buckets on four trees—and it doesn't take long to make the rounds. Amy bounds ahead, eager to lift each lid and gauge the overnight accumulation. While I dump the clear sap into buckets, Amy touches her finger to a droplet hanging from the tap, breaking the surface tension so it melts across her fingertip before she licks it clean. We've got a pretty good deal going here—a couple named Jan and Gale have all the equipment and do the boiling. They have agreed to give us half the syrup in exchange for allowing them to tap the trees. All we have to do is gather the sap and store it in two plastic barrels. When the barrels are full, it is Amy's job to call Jan and Gale and tell them so.

We've had a good stretch of weather for the sap run—warm, sunny days, freezing at night—and when Amy lifts the galvanized lids she finds most of the hanging buckets are full and capped with a crust of cloudy white ice. Sometimes when we get to the last tree, however, there is only an inch or two of frozen sap at the bottom of the bucket. Amy backs off and shakes her head at the tree sadly, as if it is an underperforming child. Then she scoots ahead to the garage, where she steadies the funnel as I decant the day's collection. The whole job gives us maybe ten minutes together, but as she skips back toward the house to begin her school day, I am hoping in memory she will recall it as much longer.

As I walk to the office the sun is warm but the wind is cold. This seasonal contrast always evokes memories of my friend Ricky. A neighbor kid who began hanging around our farm one spring when I was five or six years old, Ricky was a dark-eyed boy of about twelve who didn't seem to have friends his own age. At first, Mom says, that worried her. But Ricky and I struck up a fast friendship, aided by the fact that by country standards Ricky lived right around the corner: two flat miles from his driveway to mine. And blacktop all the way. Nothin' at all for a boy on a bike.

Not forty yards from Ricky's mailbox, a pair of corrugated culverts punched north-south through the east-west berm of Beaver Creek Road, carrying Beaver Creek itself beneath it. Two steel tubes and a middling stream might not sound like much, but as far as I was concerned, Ricky was the luckiest boy in the world.

My father's farm was all swamp and flatland. This left me easily bewitched by moving water. Water that *flowed*—that didn't just *seep*, or sit still and fester up mosquitoes—gave me Huck Finn fevers. Those first warm days coming out of winter, my siblings kept our ears cocked for the sound of trickling water. Then we'd track the trickle down and do what we could to speed the flow—kicking snow into the channel, where it melted even as it floated, or widening the channel by stomping the overhanging edges of ice, which snapped beneath our boots with a satisfying crunch. When a true thaw came, rivulets broke loose everywhere, and we spent hours gouging channels from one puddle to the next, delighting in how the dirt crumbled into the clear water, spinning mud clouds downstream to form cream-in-coffee eddies. When the sediment swept clear and again the water ran transparent, miniature rapids sparkled in the sun. If we churned the puddles to mud with our boots beforehand, the drawdown left mocha-foam striations along the shoreline. It seems the urge to control the flow of water is innate—rare is the child not born prequalified for the Army Corps of Engineers. Workable parallels are found in the urge to shovel square corners into freshly fallen snow. A man on a local radio show classifies the snow-handling fetish as a form of "space management." This is apt, but I propose freelance hydrology as a subcategory.

The water often melted faster than it could dissipate. The low spot in the middle of last year's cornfield became a pond, complete with paddling ducks; a dip in the road became a flat stretch of water hazard—a mirage that wasn't a mirage; over where the Keysey Swamp drained, the culverts submerged leaving no trace but a whirlpool that spun narrower and narrower until the gabbling swirl sucked shut and left the swamp water to rise in silence above the hummocks and muskrat houses to the very shoulder of Five Mile Road and sometimes across it so bullfrogs might laze unmolested above the centerline. Children love the idea of transformation and alternate worlds, and the delayed spring runoff transformed our landscape as completely as any fairy-tale Merlin. Once I sat very still against a white pine and watched as an early-returning mallard couple paddled within six feet of me on what in dry times was a deer trail. I was transfixed by the drake's iridescent head, so close I could see the wet shine of his eyeball. One sodden spring when I was older, the road flooded by Oscar Knipfer's place and we took the canoe over. We paddled back and forth from one blacktop shore to the other, giddy with the anomaly of it. Every summer we canoed the Red Cedar River, but for some reason it was twice as exciting to paddle above the roadway, as if we had been gifted with a magical boat.

One surreal spring day my brother John spotted a northern pike in the cow pasture. It was just a snaky little hammer handle—if we had caught it while actually fishing, we would have tossed it back without a thought. But here on the back forty, in a drainage ditch ten navigable miles from the nearest habitable lake, it was an event on par with the birth of a white buffalo. We hatched fantasies of summer afternoons spent casting Daredevils from the hay wagon. John fetched his fishing rod and managed to land the fish. Tiny as it was, he released it immediately. Then the ditchwater receded, and with it the dream of battling lunkers in the clover.

But Ricky—Ricky didn't have to wait for magic water—when the ice let loose over his way, he had himself a real honest-to-goodness *crick*, and it lasted all summer long! How I envied him this exotic proximity. During the spring of our friendship I was always eager to pedal over his way when Mom gave the OK. We'd meet at the culverts, choose one, clamber to opposite ends, then hang head-down to whoop back and forth. Our voices echoed flatly before dampening against the corrugations. We tossed pebbles at each other. They fell short and went *ploop!* or ricocheted off the ribbed steel with a compressed *ping!* Sometimes we played Poohsticks, simultaneously dropping two different-sized branches in the upstream end of the culvert before running across the road to the downstream end, hoping the stick we picked came through first.

In the first week of his fortieth lambing season my father climbed aboard a tractor (something he has done almost daily all those decades) and the knee of his trailing leg emitted a celeriac crunch, which, as it turned out, was the sound of his meniscus dismantling. He was instantly hobbled with pain, unable to bear weight, and confined to the recliner. We kids—all grown up now—take turns staying at the farm to help out. It is a chance for me to introduce Amy to a ritual that spanned my entire childhood, and I was happy yesterday when she walked with me to the barn and we discovered a sheep ready to deliver. I told her she could name the lamb.

In a practice dating back to the beginning, the lambs are named alphabetically. This was always a fun game—I remember long-gone fuzzballs named Herkimer and Knucklehead and Lillelukelani. Adherence to the alphabetical constraints was jovially strict, and led to fuzzy little creatures named X-ray and Zapata.

The ledger of record was a clipboard hung on a nail. The pencil dangled on a string. The system remains unchanged.

In the barn, Amy was eager and attentive, watching closely and asking questions as the lamb emerged. It was stillborn. She cried a little, and we talked about it. I told her that sometimes surprises are sad. Then I told her once we had a lamb born with five legs and six feet, so we named him Spyder. Two hours later another ewe went into labor, and this time Amy saw twin lambs arrive alive. As they shook themselves and tottered to life, she smiled and chattered happily, and while I have attempted a career out of overthinking things, I suspect her smile was all the wider in light of her recently acquired prior knowledge.

I have taken night duty, and when the alarm sounds at 2:00 a.m. (rousing by habit and intuition, Dad rarely requires the uncouth tool), every lazy bone in my body—to say nothing of my cotton-bound brain—assumes a specific gravity designed to drive me deeper abed. I summon the strength to rise only by conjuring the fantasy of how sweet it will feel to drift off upon my return. By the time I am dressed and downstairs, I am reanimating my childhood. On weekend nights, we kids were allowed to accompany Mom or Dad on midnight maternity rounds. There was always a feeling of anticipation in coming down to the dark kitchen and bundling up for the trek to the barn. Beyond the weak pool of the yard light, the farm was socked in darkness. Wisconsin's March is highly variable. Sometimes a soft wind was soughing in the pines, shushing through the needles and pushing the scent of melt. Sometimes the night was clear and deep-frozen. Sometimes snow was coming down. One night when big flakes were lazing past the yard light like feathers from a burst pillow, I went to check the sheep with Mom. When she held the iron gate open, I stepped through and the top of my head brushed the underside of her outstretched arm. "My goodness," she said. "Pretty soon you won't fit under there!" I felt eight feet tall and strode the rest of the path with shoulders squared.

Every trip to the lambing barn was charged with anticipation. As we looked over the flock, we listened for the sounds of labor or a newborn bleat. The animals were settled, resting like woolly boulders with their legs folded and hooves tucked beneath their bodies. If you stood in the quiet, you could hear them working their cud. Audible human mastication drives me nuts in a split second, but for some reason I find the sound of sheep chewing a soothing nocturne. An animal in distress does not bring up a cud, and all that muffled molar work—with regular pauses to swallow one bolus and bring up another—sends a subliminal message of contentment. When I was young I would climb the haystack into the rafters, then curl up and simply listen.

Tonight I hear an infantile bleat before I reach the barn, and when I straddle the fence and cross to the straw, I find a young ewe lying on her side and straining. She has one hind leg in the air like a roast turkey. There is a fresh-born lamb beside her, and as I approach, she presses out another. Arriving in a slithery amniotic gush, it plops wetly to the straw. Encircling its nose with my fingers, I milk its nostrils and mouth clear of fluid, then stand back to watch its ribs bow in and out as the first hacking breaths transpire. By the time it shakes its ears loose (this always reminds me of an accelerated version of the emergent butterfly uncrinkling its wet wings after escaping the chrysalis) I am experiencing the standard moment of marvel at how the whole deal works. The ewe has turned, snuffling and chuckling as she licks the amniotic fluid away, roughing and fluffing the tight wool curls so they can air-dry. As usual, the other sheep ignore the goings-on, with the occasional exception of the yearling ewes. Having never given birth, they sometimes sniff the lambs or the hind end of the laboring ewe curiously, their ears cocked forward in a mixture of curiosity and alarm as they nose the amniotic sac dangling like a water balloon.

Dad keeps a baby food jar filled with iodine in the barn, and I retrieve it now, removing the cap and lifting each lamb so I can thread the umbilicus into the ruby liquid. I do it the way I remember Dad doing it, clapping the jar tightly against the lamb's belly, then tipping both back simultaneously so the umbilicus gets a good soak, a practice intended to prevent navel ill. The lamb is left with a circular orange stain on its abdomen. In a week or so the umbilicus will turn to jerky and eventually drop unnoticed to the straw.

By the time I have finished with the two lambs, the ewe has gone to pushing again. I ease around behind her. I'm hoping to see a pair of soft hoof tips cradling a little lamb snoot. The hooves are there, sure enough, but they are dewclaws-up, and there is no snoot. Bad sign. These are the back legs. Breech delivery. I hustle back to the house and wake Mom. Dad has always shouldered the bulk of the lambing chores, but defers to my mother for tricky deliveries. She comes armed with delivery-room experience and delicate hands. Dad's hands are not overlarge, but they have a sausagey thickness brought on by manual labor and are therefore poorly suited for navigating obstetrical tangles.

I get back to the barn before Mom and find the ewe panting with the lamb half out—its head, shoulders and front legs still lodged in the birth canal. It appears there is no time to wait, so I grab the lamb and pull it the rest of the way out. Its head is still inside the amniotic sac. I clear the nostrils and mouth, but there is no breath. I give a couple of pushes on the ribs and dangle the lamb by its back legs, which looks drastic but allows fluid to drain from the air passages. When I place the lamb on the straw, its flanks flutter, and then I hear the familiar crackle of air working into the lungs. Shoot, the little feller's off and running. Mom arrives. Minutes later the lamb gives a high-pitched bleat, and I am just plumb happy.

We stand and observe. Let the new family get to know one another. Mom kneels behind the sheep, checks inside to rule out quadruplets. Nothing. The ewe's long push is over. Using another trick my father taught me, I guide the sheep to the pen by dangling the third lamb in my hands while slowly backing across the barn and into the small square pen. It takes a while—the mother wants to dart back and forth between lambs, so I carry two and Mom the other—but soon they are ensconced, the two oldest lambs already stumbling about in their jabby-stabby knock-kneed way. The breech lamb is worn out. After watching the first two lambs suckle, we try to help him latch on, but he's tuckered. Dad says the emerging thinking is that immediate nursing isn't as necessary as previously thought, so we'll leave and let the family settle. Over the course of the coming day we'll keep an eye on the little guy. Make sure he learns how to get his dinner. Mom jots the ewe's ear tag number and the sex of each lamb on the clipboard, but we leave the name spaces blank. Amy can name them in the morning. We return to the house. The frozen air is bell-jar still. The sky is deep black, the stars pressing down brilliantly all around, and I am reminded that we are not beneath the constellations, but among them.

When I was a young boy and accompanied Dad to do the checks, once the lambs were dipped and penned and the clipboard record updated, and we were back in the house, he would disappear into the cellar and come back up with a mason jar of canned dewberries. We'd have a bowl. The dewberries were sweet, their dark red juice reminding me of the iodine in the baby food jar. Tonight, no dewberries. Mom is off to bed and I cross to the kitchen sink, where I begin to scrub my hands. I am soaping up when I realize my wedding ring is missing. It must have come off during the delivery, when my hands were slick with amniotic fluid. I grab a flashlight, retrace my steps, and spend a good hour diligently searching the straw. Nothing. Later some wisenheimer asks if I checked inside the ewe. Well, no. But perhaps next year we can expect a little miracle lamb born with a golden band around one ear.

When Dad hurt his knee, he went to the doctor's office using his shepherd's crook as a cane. The crook came to his shoulder so he kinda hung off it with both hands and hobbled along. If I was a twelve-year-old I would have been mortified at the image. In my forties I shake my head but feel secretly happy that unusual fellow is my father. He's not sure if he's going to lamb another year. If he does sell the sheep, it will be a big deal. He was gentle with all of his animals, but I suspect the sheep speak to him on a level the cows never did.

One day I asked him if he had sheep because of their biblical significance. "I've had people ask that before," he said. "That's part of it . . ." But then he doesn't elaborate. He is quiet for a minute, apparently reflecting on forty years gone by. "The sheep were always good to us," he says, finally. "We couldn't make a living on them, but we made a quarter, or a half. A lot of years, they were the difference."

Late March, and out of nowhere, we get an eighty-degree day. The winter has been low on snow in the first place, but with this absurd burst of heat, even the holdout patches are draining away. I take advantage of the temperature to begin establishing a pig-pen in the overgrowth just downhill from my office. Somewhere in the wooded slash of valley below, a murder of crows calls as if spring is full-blown, but all the caw-cawing ricochets through leafless trees with an extra layer of reverberation that betrays the true season.

I'll scub the pigpen together as best I can. Until a few decades ago this was a working dairy farm, and the patch I've chosen for the pigs seems to be roughly where the barn once stood. The remnants of a paddock—weathered planks spiked to railroad ties sunk vertically in the earth—still stand along one edge of a long concrete slab that appears to have functioned as a feed bunk. The rest of the fencing is mostly teetering or collapsed. Several of the wooden posts are rotted off at ground level. But enough of one corner remains intact that I believe I can close it off and create space adequate to contain a pair of pigs.

One little farmstead, and there is so much to learn. In the odd available moment, I make explorations. On a previous nose-around I discovered several steel cattle panels in the brush, and today I go about extracting them. It's a sweat-making, itchy task. This early in the year there are no nettles or poison ivy, but the panels are trapped in skeins of wild cucumber and woody twists of grapevine deep within six-foot-tall banks of burdock. Furthermore, a few tenacious staples still hold fast to the posts. Diving in with my fencing pliers, I cut and yank and tug. I have always regarded brute force as an acceptable first option. Eventually I free the first panel. When I drag it out into the open I'm speckled with dirt and duff, and my shirt is so gnarled with burdock burrs it looks as if I've been swarmed by a horde of miniature hedgehogs. I dump the panel on the dead grass and burrow in after the next one. Once I get off my lard and rolling, I am a sucker for grunt work, where the most difficult problems are solved by getting a better grip and putting more heave in your ho. While the bones and meat wrassle, the mind is free to sort and ponder.

The project goes well. In just under sixty minutes I have six panels flat on the dead grass of the paddock. Brand-new, these panels tally up around sixteen bucks apiece at Farm & Fleet. I congratulate myself on being self-employed at right around a hundred bucks an hour. Sadly, this fiscal spike is what your statisticians call an "outlier" and is unlikely to skew long-term results.

With the panels at hand, I set about fashioning the pen. In the corner I have chosen, several panels are still upright and attached to relatively sturdy posts. A twist of wire here, a staple driven there, and they're pig-worthy. In another spot a gaping hole has been torn in the panel—they are made from heavy-gauge welded wire, so it must have been someone careless with a front-end loader. Back in the brush I find a short section of panel that covers the gap to near perfection. I wire it in place with short lengths of electric-fence wire snipped from a tangle I found twisted around a decrepit plastic insulator tacked to one of the railroad ties. When the patch is in place, there is still a small vertical gap about the width of a piglet. Following another rusty strand of electric-fence wire into a patch of blackberry stalks, I find a deformed electric-fence post. It is bent in such a manner that I am able to weave it in and out of the panel on either side of the gap to form a rebar suture obstructing the hole. As much as I would like to have a spotless, squared-up operation, I need only look around the rest of my life to know it ain't likely, and further-more, having recently priced fencing supplies while resting in the bathroom with the latest Farm & Fleet catalog, I am getting hooked on the idea of salvage in terms of budget enhancement. I wiggle-waggle a steel post ($3.25 new) loose from the old fence line and drive it to extend the reach of the pen, the thaw having been such that the earth here is soft.

I work well into the afternoon. By the time I decide to hang it up, I have formed a three-quarters-sided enclosure and am surprised to find my shoulders sunburned. I am snugging the last twist of wire down tight when a fly buzzes past. The sound is out of place for the season. Perhaps the world is changing. But there could be snow tomorrow. The fly should not get his hopes up.

The culverts where Ricky and I played still remain. The tubes are sunk deep and solid. Should you run that stretch of Beaver Creek Road, you will detect no change in the hum of your tires as you pass across the blacktop above. Depending how fast you're flying, you may fail to even note you've crossed the eponymous creek. When Ricky and I heard cars coming we would scramble off the culverts and hunker in the ditch, below the sight line and hidden in the grass. No driver ever spotted us.

In winter the ditches that fed into Beaver Creek were frozen solid, and in the summer they clogged and went to sluggardly soup, but during the melt the water moved with a pristine chuckle. Once while we were down out of the wind and the sun grew hot on our backs, Ricky knelt and drank deeply from the ditchwater, telling me to do the same. "Go ahead," he said. "This is how you survive." The water ran so clear above the tan sand, you could spot the individual grains. It's *pure*, Ricky said of the water. You can *see* it's pure. And so I drank too, and deeply. Later, when my mother heard, she told me about giardia and protozoa. To say nothing of dead skunks and Atrazine.

Ricky had an army surplus shovel. He carried it everywhere. It had a stocky wooden handle like a billy club, and you could fold the spade back flat to make it even more compact. On the opposite side of the spade was a pick that folded out at a right angle. Ricky and I were forever digging forts and hideouts. I recall a buried culvert with a trapdoor, but surely this was one of Ricky's dreams and not a reality. Although the memory is precise enough—I see it somewhere near the machine shed and up against a row of Norway pines—that if Ricky were here, perhaps he could tell me that indeed it was so.

Once Ricky invited me to lunch. I called Mom and she said it was OK. I remember two things: there was a partially assembled large-block engine on the floor in the living room, and we had runny eggs. The runny eggs were a novelty for me. No offense to my mother, but I had never seen a fried egg presented in any manner deviating very far from vulcanized. I remember now that Ricky's brother Alan was at the table with us. Alan wore old army jackets. Several years later, he killed a man. I read about it in the

Chippewa Herald-Telegram. There was trouble involving Alan's and Ricky's sister. Alan put four bullets in the man's chest and went to prison.

I have always thought of my friendship with Ricky as spanning several years, but having gone back to look at photographs and having spoken with my mom, I realize the friendship was at its peak that single spring and was probably over by autumn. We never had a falling-out, and Ricky never told me to get lost. There was just a slow dissolve to other arrangements. I think it probably had to do with our age difference. Once the summer ended and we returned to school, Ricky was headed around the bend to high school, while I remained in the grade school wing. From his perspective I suspect the social gap was insupportable. My last memory of the young Ricky is sad. We were on the school bus, headed home. I was in the seat ahead of Ricky, sitting sideways so I could talk to him. One of the rough boys, a stocky football player, barged up the aisle and demanded that Ricky move from the window. When he didn't move fast enough, the bigger boy dove into the seat and landed on him, heavily. Ricky was holding that army shovel on his lap, and when the lunkhead dropped on him, the metal edge of the blade drove into Ricky's thigh. It didn't break the skin, but it hurt terribly. So terribly that Ricky burst into tears, and the big boy laughed at him. I remember trembling angrily at the big boy but being too small to do anything about it, and ashamed that Ricky—my older friend, my hero from the ditches—should have to cry in front of me.

I work on the pigpen two days in a row. My brother John said I could have his old hog feeder, so I run up north to retrieve it, using the trip as an excuse to swing by Farm & Fleet, which, as a guy likes to say, is "right on the way."

It usually is.

My favorite thing about Farm & Fleet has always been the smell of fresh tires, but the livestock corner holds its own with a potpourri of alfalfa cubes, Terramycin crumbles, horse vitamins, and the malty sweet scent of milk replacer mix. When I pass the stacks of rough paper sacks containing calf starter, the smell of moist grain and molasses reminds me that we ate it by the handful when we were kids. Dad called it "calf candy," and it wasn't bad. Most of it was fortified with antibiotics, so we rarely got the scours. Over in the feeder section, where the galvanized grain scoops and hay racks are for sale, I pick up a heavy rubber pan for feeding slop. Then, with a rough idea in mind of how I might construct a watering system for the pigs, I also load the cart with an adjustable spring-loaded spigot, some tubing, and a bagful of pipe clamps and plastic reducers. As I head for the checkout I pass a rampart of salt blocks piled on pallets just the way Dad used to store them. Each cube is roughly the size of a car battery. We used to drape ourselves over the stack side by side and lick the blocks. The salt was coarse against our tongues, like licking fine-grain sandpaper. If we kept at it too long, our tongues got raw. Dad always got the reddish brown blocks with trace minerals—there were no goiters in this family. I'd give the blocks a lick now, but I don't want to freak out the guy watching the security cameras.

Back home, I rig the waterer, using a plastic barrel I got from my friend Mills. I mount the waterer on a hastily arranged tripod, and the elevation is sufficient so that the water runs down the hose and out the spigot. I am not much of a talent toolwise, but this has gone well, so when I am done I stand back and give it the classic male postproject lookover and am satisfied. After two days shirtless in the freak March sun, I am deeply burned. This is medically foolish, but here up north we worship the sun in big gulps.

A week later, and it is a gray, mist-spitting day. The warm weather has continued, with a moderation from ridiculous to mildly unseasonable. Amy and I are stacking firewood. She is expected to pitch in as standard procedure, but this time it's a bit of a shanghai, as she is being compelled to stay home and work while Anneliese runs errands in town. This is the promised consequence of a recent in-store meltdown. She is weepy at the get-go, but then as so often happens if one maintains one's parental resolve and resists either cave-in or eruption, about twenty minutes in we are happily chatting, and by the time we stack a half-cord, she is flat-out jabbering. "I'm glad I didn't go to town!" she exuberates at one point, and it briefly strikes me that this calls into question the very efficacy of the punishment, but I abandon this train of thought as unproductive. Sweating as I always do when I do anything more physical than lift a pen, I tell her about my friend Frank, whose father taught him that firewood warms you twice—once when you split and stack it, and once when you burn it. I predict by the time Amy is nine, "Firewood warms you twice" will make her list of Top Five Phrases Most Likely to Make Me Roll My Eyes at the Old Guy. Somewhere from the piney draw below us comes a pheasant's sore-throated squawk. Of course we cannot know, but we wink at each other, assuming it's Mister Big Shot in hot pursuit.

We work for two hours. Then we spend a little time picking up the usual bits of yard garbage revealed when the snow retreats. All the bare ground reminds me that I have promised Anneliese I will make a cold frame for the garden, so I wander around the sheds rustling up scrap lumber and an old storm window, a box of drywall screws, and two rusty hinges. In about twenty minutes I clatter together what could pass for the junior high shop project of a three-fingered monkey, but then I cut myself some slack and declare it evocative of a sculpture I once stumbled across in a stairwell at the 2002 Whitney Biennial in New York City. Amy and I scratch up a patch of ground near the spot where Anneliese's mother had last year's garden, and then we plant lettuce, radishes, carrots, and some parsley. It's a rush job, and we'll see how it goes. The ground is heavily threaded with earthworms, and we discover a stand of garlic shoots already four inches tall.

I set Amy free then. She runs off to play with Fritz the Dog, a German shepherd and one of two dogs we are sitting for friends. I walk across the yard to store my tools in the old granary. The day is still misty, and twinkling beads of precipitation hang from the underside of the apple tree's slenderest branches. Down in the valley the pheasant is still squawking. In the yard, a male mourning dove drops groundward and lands just behind his female companion. He hops toward her tail, then flutters just above her until she flits briefly ahead. He follows, hops, and flutters again. A lighter-than-air tumbling act, they hopscotch each other all across the lawn until I come too near and they spook into a wing-whistling takeoff. At first burst, white bars flash from beneath their gray-brown wings; then they swoop to roost atop the granary, settling nervously atop the ridge cap, dipping their heads and side-glancing my approach.

Inside the disused wire-frame corncrib just beyond the granary door, two juncos are chasing each other in abbreviated figure eights. Between flights the juncos drop to the circular concrete floor of the crib and scamper their own fluttery do-si-do, the rain-slick slab an impressionist mirror reflecting their jitter-bug. Inside the granary I see barn swallows daubing a nest in the rafters. We live in a time when earth cycles are in question. I look at the yard—frost-free and soaked, already with an undertone of green—and the trees with their buds preternaturally frayed, and I think there is certainly evidence for discussion, but then I look at the evidence of all the birds this morning, and it is clear some cycles remain resolutely intact. Gray and wet it may be, but the birds are sunny in love.

There are the usual deadlines, so I climb the path back to the office. I am squeezing all the pig-penning, wood-splitting, cold-framing, and daughter-consequencing in between the desk and road time that pays the bills, and looking around me at all the relentless evidence of time and seasons passing, I hear the little voice telling me that a guy ought to pare down. We are a breathless society. I love what I do and am grateful to do it, but I am hooked into short-winded cycles of my own, and a simple move to the country does not stop the clocks. It strikes me that this morning's chores should have ended not with me checking the time and switching the computer on but with Amy and me taking a long stroll into the valley, to learn the land together. This room above the garage gives me a wide view of the place, and I can see her in the yard beside the granary, squatting in her pink rubber boots with her arms wrapped around her shins, nose to nose with Fritz the Dog, who is currently chewing on a dead rabbit. As I wait for the computer to boot, I watch Amy lean in for a better look as Fritz gnaws away at the rabbit's hind leg. She turns her head this way and that, studying the carcass from every angle as the dog grinds through hide and muscle, working the skull back to his molars so he can crack it and taste the brains. When he curls his lip and pulls at the guts, Amy leans in so close I expect her to topple. With no other dog to compete, Fritz is eating leisurely. A good fifteen minutes pass before he is nosing the final morsel—a front paw—around in the grass between his own front paws, and Amy is still squatted there, transfixed. Anneliese and I constantly second-guess ourselves as parents. We wonder if we are sometimes too strict regarding issues such as the enforced wood-stacking. We wonder about the effects of me being on the road as much as I am. We wonder if we are projecting our idea of country living too heavily on her childhood. We wonder whether we are cheating her of our own happy public school experiences by homeschooling her. Whatever the case, I look at that little girl out there, now on all fours to watch the unlucky rabbit's foot disappear down the dog's gullet, and I think, well, it's not like there's nothing to do.

Ricky died not so long ago. His obituary was a surprise, even thirty years down the road. There had been no contact, although I saw him a couple times in his truck, an old L-model International he had converted to four-wheel drive. I was in college at the time, and Ricky was helping Dad with odd jobs and logging. We said hello, but he was whip-thin and furtive, and the conversation didn't go anywhere. Later I read in the local weekly that there was trouble at the grocery store and when the cops found Ricky walking afterward he had a gun, but he gave up quietly and went to jail. And then he was dead—not young, but too soon, and alone in a small apartment. I never asked how. I made it graveside and stood in the cold wind while one of his friends put a boom box on the headstone and played a song I should have written down because now I can't remember. His daughter was there, with the same dark eyes I remembered from Ricky the boy. Hers were reddened with mourning, but she was wearing an army dress uniform, and you could see her standing tall because she knew it would have made her daddy proud. Afterward we went to the McDonald's right across the street from the cemetery and we all had some coffee like Ricky had in that same McDonald's every day for the last several years. Maybe he'd seen it coming clear back when we were kids. He had some sadness on him. It came built in.

My friends Andy and Wendy helped me put together a video essay about Ricky and the culverts for Wisconsin Public Television. Then a magazine asked me to write about my favorite place in the world. The question is unanswerable (there is a mountain in Carbon County, Wyoming, that pulls at me like the moon; there is a pine tree near here that fits the curve of my back; once I stood in a ruined Welsh castle and felt a thousand years old), but I chose the culverts for that, too. Ricky's daughter saw the television piece and wrote me a letter. When I started the magazine essay, I wanted to reread the letter, so I dug through the piles on my desk until I found it. When I pulled the folded paper sheets from the envelope, a pair of photographs fell out. They were of Ricky—when I opened the letter the first time they had stuck inside and I hadn't seen them. In rough notes toward the essay I had mentioned Ricky's dark eyes but wondered if I was recalling them accurately, as memories have a way of conforming to our stories the more we tell them. But there in those photos—one of Ricky as a young man and one of him older, from the years I didn't know him—were the very eyes memory conjured. I must restrain my speculation; there was so much more to this man than my few stories predicated on our childhood days, the odd newspaper clipping, and a funeral. But looking at those eyes now, I think Ricky knew early on he wasn't suited for this world. I think he carried that army shovel figuring if worse came to worst he could at the very least dig in. Thing is, we never did finish any of our hideouts. I think Ricky died still digging.

You learn not to pretty these things up. You learn to take them as they are. I go to the culverts one day and just sit quiet. Two steel tubes and a halfhearted creek: I guess I could do better for a favorite place. But grandeur is for postcard trips. For the long haul, I want the click and trickle of flat water moving, the shelter of the grass, a road close to home. The chance to slip from sight at the sound of motors. I throw a pebble for Ricky, but I'm not looking for angels in the tag alders. I just watch the creek flow from beneath me and out of sight around the bend. When I was a kid I yearned to follow that water—on a raft, in a canoe, maybe simply barefoot with a stick. Now I just dangle my boots and let the cold spring air make my nose run, and I watch Beaver Creek slide smooth and quiet until it reoccurs to me that the world is constantly trying to bring everything level.

I have gone in to Eau Claire to hang out with some of my fire-fighter pals (including my friend Mills) at Station #5 when I get the call from Anneliese. "I'm having contractions," she says. "I'm not sure this is it, but they seem to be getting stronger."

"Are you saying I should come home?" I ask. Specific instructions work best.

"Yes."

The crew is just making supper, so they send me on my way with a tin of homemade lasagna. Someone wrote "Good Luck!" on the container. I thought that was nice. Driving home through the lowering light, I don't know what to think. In fact I am numb to the idea of what's happening. When I get home, I stow the car in the garage and walk in to find Anneliese. She's on the couch. I hug her, ask her how she's feeling. The contractions are steady, she says. Cripes, I think, here we go. I get a watch and time a few. Then I call my mom. And then Leah, the midwife.

Leah arrives around 8:00 p.m. Her student and assistant arrive shortly after. When Mom and Anneliese's friend Jaci join us, I look around at all the women and I'm grateful, but I'm also wishing my buddy Mills was available. We have a twenty-year history now, having met when he was a medical first responder and I was a freshly minted nurse and EMT. These days he is a full-time firefighter and paramedic and I am a medical first responder. Now that I am back down in this area, we occasionally respond to the same emergency calls again. Only now, when we meet on scene, he is the one in charge. It's not the first time I've experienced responsibility role reversal—truth is, I enjoy it. In his career, Mills has delivered six babies, so a while back when I realized the home delivery team was trending all-female I asked Mills if he would be my doula. *"Y'wha-wha?"* he said. Only partially tongue in cheek, I explained that a doula provides physical and emotional support through the birth process. He beamed and accepted. Unfortunately, tonight he's pulling a twenty-four-hour shift back at Station #5 and won't be off until tomorrow morning.

We move upstairs to the bedroom so Leah can examine Anneliese. For the first time I notice Anneliese is trembling. I've never seen her so vulnerable. I hold her hand and she squeezes back and it hits me how powerful this is going to be, and then Leah says, "You're only two centimeters." Quite a ways to go yet, then. Anticipating the long night ahead, Leah and her helpers go into a back bedroom to sleep. Leah recommends that Anneliese try to do the same, but Anneliese is too nerved up, so we go back downstairs. I stoke the woodstove, and we time some more contractions. Then Jaci takes some goofy pictures, including one of me staring at Anneliese's bare belly with a look of bewilderment. I really don't have to dig too deep for motivation.

Many years ago when we burned the old feed mill in New Auburn, I was allowed to rescue the blackboard where the managers used to update feed prices. It's made of enameled steel and reads CO-OP FEED—ANIMAL HEALTH across the top. It hung in my New Auburn kitchen for years. When we moved into the Fall Creek farmhouse, I hung it from a nail in the kitchen here. Jaci has been using the blackboard to log contractions. Beneath the times you can still make out faint renderings of the price of cracked corn and sunflower seeds. It's nice, sitting there on our old couch with a good fire going in the stove and my mother off to the side knitting, her aluminum needles clicking softly in the yarn as Jaci keeps time.

And then it all stops. The contractions fade, then cease. Hoping for a kick-start, Anneliese and I go for a walk. Outside, the wind is wintry cold, and oak leaves skitter across the driveway. The warm spate is over, and it feels more like autumn than spring. We walk out the drive and down to the mailbox, then back up the drive and out the ridge, where we stand quiet for a while. The moon glows behind a thin veil of clouds, shedding just enough glow so that we may see the general shape of the land. I hold Anneliese close, her cheek cool against mine. I can feel her trembling still, but I don't feel very sheltering or strong. Sometimes I don't make much of a grown-up. I'm a little boy who prefers to shape his stories just so.

It is nearly midnight when we head back inside. Leah rises to check Anneliese again. Still two centimeters, and the contractions haven't returned. "Get some sleep," says Leah. "Rest, in case things start again." She goes back upstairs to sleep some more herself. Anneliese and I climb the stairs. Lying in bed in the dark, I remember the Friday night in high school when we got all revved up for kickoff and then the ball blew off the tee. I admit the analogy has limitations and may not translate across the gender divide. In any case, I have the rare good sense to keep it to myself. I can feel the disappointment and frustration in the way Anneliese lays beside me. Eventually we sleep.

In the morning everyone is gone.

On the chalkboard Jaci has erased the contraction times and written:

THURSDAY EVENING
SHOW
POSTPONED
Due to
Stage Fright

There we were with that stretch of glorious and fraudulent weather, and now we are back to stinging ears and snow on the ground and foolish jump-start robins shivering in the maple trees. Many of the early-breaking buds are frost-burned black. One of the maples flanking the path to my office has a broken limb, and an icicle of sap hangs from the fractured wood. The run of warm weather brought an abrupt end to the sap run, and we pulled the taps. Once the trees bud out, the clear sap turns faint red and bitter—professional sugarers say the sap has "gone buddy." Amy and Anneliese went to observe the boil-down with Jan and Gale, and now we have a gallon and a half of maple syrup in the pantry as well as a few maple sugar candies in the freezer—technically the first food from our new patch of land.

So it's cold again, but the earth is turning. Nighttimes it's been dropping to the teens, and the muddy spot on the office footpath is coated with ice, but it fractures easily when I step on it, and mud oozes up through the cracks. Down on the woodpile sits a mason jar. The day we stacked wood Amy noticed me sweating, and, unbidden, filled the jar with water and brought it to me. I drank it down to an inch from the bottom and set it atop the stack, where it sat at such an angle that now the base is filled by a lopsided puck of ice. I see the glass there on the split oak and turn immediately maudlin, blindsided by the idea that the jar and the water are representative of how the most fluid, workaday moments become fixed in sweet irretrievable history in the very instant of their occurring.

I have promised Anneliese that when the baby comes I will spend an entire week with her and the new child, returning no phone calls, answering no e-mails, working toward no deadlines. In the meantime, I am churning away as usual, constantly rearranging the days into an endless chain of last-minutes. I see that glass as an emblem of placidity surrounded by the snarl of my subsequent overbooked peregrinations and hustle. Long ago, I think, my daughter drew water and brought it to me. A grand thing in its simplicity. I lift the jar, then replace it, suddenly convinced that it covers a hole where all the time drains away.

Later in the day Mister Big Shot appears in the yard. At his side, a girl bird. He struts beside her as if a tail ain't nothin' but a drag. I think of me beside my wife, and then I think, even us bald guys get lucky sometimes.

Just as when Anneliese had her spate of Braxton-Hicks contractions nearly three months ago, I kept obsessively checking the baby's heartbeat after the night she thought she was giving birth. And every now and then for the next several days I keep asking Anneliese if the baby is still kicking as before. She assures me it is. After all the ramp-up with no payoff, we've been left a bit adrift. The bright blue birthing tub stands at the top of the stairs, the water perfectly still. We walk around it.

A few days after the fact, I talk to Albert Frost, an old-timer from up by the home farm. Albert is in his nineties, his wife dead some ten years now, but still lives on his farm within sight of the culverts where Ricky and I used to play. Albert was always skinny as a crow's leg (my brothers call him "Fat Albert" and grin) and nowadays he uses a cane, but he has stayed on the home place and stubbornly fends for himself. I tell him we are waiting on a baby. Tell him about the false start. He chuckles. "When my first boy was born, there was a storm coming," he says. "They claim a big storm will bring it on.

"They had seven babies at the hospital that day. My kid was born at eight in the morning. By noon I still hadn't seen him. So I asked the nurse, and she held him up behind the glass.

"Homeliest little fart you've ever seen. I was pretty disappointed. But I thought, 'Well, he's healthy. I better not complain.'

"Then I heard the nurse saying, 'What's your name?' I told her, and she said, 'This one isn't yours,' and she held up another one." He laughs. Like it was yesterday.

"Yeah, but Albert," I say, "did that one look any better?"

He's still chuckling. "Well, *I* thought so," he says, "but I suppose I was prejudiced."

CHAPTER 5

Across the valley, the bare-bone tree line is thickening. The maple leaves are fit to bust but holding fast, this year's greenery still clasped in a tight fetal furl. The bud scales are dark red, infusing the canopy with a rubrous blush, shrouding the hills all smoky maroon. It is mid-afternoon, sunny, and still. I hear sparrows.

There is a baby on my lap.

Ten days have passed since the false alarm. It has been tough on Anneliese, going right to the precipice only to have her body shut down and scuttle the whole production. The sleeplessness returned tenfold, and with it the doubt, the brittle emotions, and the desperate weariness. She is occupied above all with the desire to get the baby born.

The morning after Easter I am at my desk above the garage when I see her pass by the window. She comes through the door and sits wide-legged and heavy in the saggy green chair. "I think maybe it's happening," she says. Apparently she had been up at 2:00 a.m., timing contractions while lying on the floor beside our bed. At some point they faded and she climbed back in bed and went to sleep. Ever helpful, I slept through the whole thing. Now the contractions have returned. "They're strong enough that I have to stop and wait them out," said Anneliese. We chat a while. A handful of contractions come and go. Then, as Anneliese stands to leave, a big one hits. She bends over, cradling her belly with one arm. She grimaces and blows through pursed lips. When the contraction passes, she returns to the house, and I phone Leah the midwife. We talk it around a while, me not wanting to pull the alarm early again, but Leah says it sounds like she should head our way, especially since she has a ways to drive. When I get to the house I find Anneliese on the sofa, gripped by another contraction. Her mother, Donna—who has been visiting more or less on standby—is at her side.

Shortly, Jaci arrives. She and Donna take Anneliese for a walk along the ridge. When the three of them return, the contractions are coming apace and Anneliese has to stop whatever she is doing to breathe through them. She says it helps if I rub her back, and while I am doing this, I notice Amy hovering around the edge of everything. She is beginning to look apprehensive. Since the time we began to plan for a home birth, Anneliese and I have talked with Amy several times about whether or not she would like to be present for the delivery. I've been torn about it from the beginning. I'm all for it if she wishes, but I also can't see any reason she should be compelled to stay if she is disturbed at the sight of her mother in distress. All along she has been saying yes, but right now her eyes are a little too wide. We talk it over again now, and Amy says she wants to come upstairs with us when it is time, but I also discuss it with Donna and she agrees to take Amy out of sight and earshot if she so requests. For now Anneliese and Amy go outside together and sit in the hot tub beside the deck.

When Leah arrives she goes to the deck and visits with Anneliese. I'm off muddling around, checking the water in the birthing tub, looking for my swimsuit, wondering if I should sneak one more high-speed cram session with *Emergency Care in the Streets*. Out on the deck, Leah tells Anneliese, "Well, we might as well check you."

"If I'm at four centimeters, I don't even want to know," says Anneliese, wrapping herself in a towel and walking into the house.

I kneel beside Anneliese, holding her hand as Leah performs the examination. Leah's eyebrows shoot up and her eyes widen. I'm startled, thinking something is wrong. "You're at six or seven," says Leah. "Looks like you're on your way!" Anneliese beams, and yet at the same time I see an edge of determination set in, as if she is saying, OK—let's go here.

I call Mills. He's puttering in his wood shop. Looks like this is it, I tell him. "I'm on my way," he says.

After that, I go into what can most charitably be described as a cotton-headed sleepwalk. The feeling in the pit of my stomach is like unrisen bread dough. Not dread, exactly, but *reality*. I see Mom's car in the driveway. Perhaps as a means of avoidance, I become obsessed with preparing the birthing tub. I remove the cover and stow it. Return to check the water temperature. Decide the level is a little low and go down to get a bucket of water from the laundry room. With the midwife and my mother at her elbow, Anneliese moves upstairs.

Mills arrives. He is is wearing camo pants, black Crocs, and a ball cap. It's a relief to see a scruffy male. And it doesn't hurt to think of those six babies he's delivered under all conditions. He has a batch of newspapers and several copies of the *Tradin' Post* under one arm and his Big Gulp mug of water in hand. "Go ahead and hang out in the office," I tell him. "If I there's trouble, I'll shoot off a flare." I'm hoping my bravado doesn't sound as tinny to him as it does to me.

Back in the house I join Amy in the bedroom beside Anneliese. My mother is at the foot of the bed and Donna is across the room at the window. Now the contractions have become painful. Anneliese is quiet, but her face contorts with focus as she breathes through them. It helps when I press against her lower back just like the nice lady taught us downstairs in the living room the day we got the giggles. Between contractions Anneliese smiles at Amy and speaks soothingly. Amy smiles bravely, but I sense she is ready to crumple and run.

When the water breaks right at the peak of the next contraction, it catches Anneliese off guard. "Oh!" she exclaims. Frightened by the rush of fluid and the pitch of her mother's voice, Amy begins to cry. Donna scoops her up and takes her downstairs. I follow, and taking Amy out on the deck, I hold her in my arms and explain what has happened. I tell her what it means that the water broke, and remind her of the times we talked about it before, and that it is good that it has happened. I tell her that it is very hard for Mommy to give birth, but that Mommy is very happy. She rubs at her eyes, and nods, and hugs my neck, and I tell her it's OK if she would rather do something else for a while. She nods again, and when I am back upstairs I hear the clang of the empty steel trailer bouncing behind the four-wheeler as Donna takes Amy out to gather firewood.

I check back with Anneliese, then take the bucket back downstairs into the laundry room and begin filling it again. I'm running the water over my hand, adjusting the temperature, when the apprentice pokes her head through the door. "I really think you need to get up there," she says. I follow obediently with my pail of water.

Anneliese has gotten more uncomfortable and has decided to move to the tub. Leah and I help her in. A terrific contraction catches her with one foot in and one foot out, and we're hung up for a while. "I don't think I can make it in," Anneliese says, and I get panicky visions of the baby dropping out right there. Then the contraction wanes and she settles into the tub. I scoot (I have now cranked it up a notch) into the closet to change into swimming trunks in case I have to crawl in the tub. Then I come out and position myself behind Anneliese to massage her shoulders and let her rest her head against my chest between contractions, which are growing in strength and frequency. Leah is coaching calmly, Mom is watching from the landing of the stairs, and the apprentice is standing by.

"Why don't you come around front now, Mike," says Leah. Mom takes my place at Anneliese's shoulders. I'm feeling relatively calm, and thinking clearly enough to actually recall something from one of the books Anneliese had me read: *OK, yes, this is the part where it is important for me to maintain eye contact with Anneliese, to pay attention to her breathing, and I should . . .*

"Would you like to hold the baby's head?"

Fhuzawhaaa?!?!

But yes! There it is, the head crowning already. Leah's hands are strong and steady as she guides mine down to the slimy little skullcap that fits perfectly to my palm. Leah is coaching Anneliese to push between contractions, but the coaching doesn't last for long, because the head is rapidly emerging, and what I will remember forever is the *fierceness* of my beautiful wife as she made that final push, her teeth set, her animal cry and her blue, blue eyes locked dead onto mine and suddenly the baby was out and in my hands beneath the water.

From my reading about water births, I know there is no rush, but looking down through the water at the creature in my hands, instinct takes over and I try to lift it to air. Anneliese's eyes pop wide. "Hey, that's still *attached*," she says. The baby is still submerged. I hear Leah's voice, calmer, gentler: "It's fine, it'll be OK, just wait," and she presses my hands back down.

I'm only half OK with this. I trust Leah and her apprentice, and I know Anneliese is at ease, but there is a mighty strong part of me that wants that baby above the waves and drawing oxygen. When Leah finally nods, I hand the baby—more carefully this time, with an eye to the cord—up to Anneliese, and she takes it to her breast with an ineffable motherly *oh!* and then the slimy blue bundle cuts loose with a wail in the outraged key of life and I feel a flush of relief.

Donna brought Amy up the stairs just as the baby cleared the water. Now Anneliese turns the infant to verify what she has felt, and yes, we have a little girl. "You've got a little sister, Amy," she says, and any trace of trepidation washes away in the wide smile that breaks across Amy's face. I'm tickled about this. For months Amy has been saying she wants a sister, and then very dutifully tacking on, "but a brother would be good too." So it is wonderful to give her the gift of a sister.

I move back around behind Anneliese and now we are all gathered: Leah kneeling beside the tub in her scrub top and gloved hands, the apprentice also in gloves and wearing her *Midwifery Today* T-shirt, my mom standing smiling in her long skirt, Donna and Jaci in the stairwell leaning over the rail, Amy still in her wood-gathering sweatshirt with one arm around me, and there at the center of it all, Anneliese holding the baby to her breast. Sunlight is streaming through the window, unimaginably bright to the baby I suppose, even behind her squeezed-tight eyes.

We linger around the tub. Donna kneels beside me and greets her new granddaughter. Mom tells the story of how when I was born I scuffed my nose during passage, and when the nurse—a battleship matron—dangled me for all to see, Mom took one look at my abraded schnozz, laughed, and said, "Rudolph the Red-Nosed Reindeer!" at which point the matron drew me back protectively and gave Mom a stern talking-to.

I palpate the strong pulse of the cord. Jaci circles the tub taking photographs. In this digital age we get to check them out right away, and I am surprised to see Anneliese and I both have flushed, rosy red cheeks. Amy watches closely as the baby tries to suckle, and I am happily flabbergasted at the sight of the infant's lips twisting reflexively toward the target. In time Leah clamps the cord, and I am surprised at the tough feel of it when the shears cut through.

When Anneliese is ready, we move to the bedroom. It is a walk of maybe ten feet. The midwife's apprentice has the baby wrapped in a cloth and dangling from a spring scale. She's squinting at the markings and trying to get a reading. "Eight pounds? Or eight pounds one ounce?" I jump right in: "Make it eight!" Round numbers, you see. Easier to remember. Anneliese's sister Kira has arrived, and joins my mother, Donna, and Jaci in the room. Amy is sprawled on the bed, head propped, watching the midwife rewrap the baby. I wonder what Amy will take from this moment, cupped as she is in a strong half circle of women observing new life.

I hike out to update Mills. He's in the saggy green chair nursing his Big Gulp, reading the papers grandpa-style, each section neatly folded and stacked beside the chair as it's finished. I suppose he can tell just from my face that things have gone fine, but I have to say so anyway.

"Everything's 10–2," I announce. Old-school emergency radio code. We learned it together twenty years ago. "10–2" means everyone's safe and everything's OK.

If Mills had grinned any bigger he'd have sprained his ears. He stood up, grabbed my hand, and shook it good.

He walks to the house with me. Climbs the stairs, says a quick, gentle hello to Anneliese, peeks at the baby, and takes his leave. At the top of the stairs, he stops. "Need anything?" "Nope," I say. And away he goes. Among the bedrock gifts of time are friendships expressible in five syllables or less.

When Anneliese gave birth to Amy, there were no afterglow moments—torn and hemorrhaging, she went straight to surgery. Today she has a small tear but instead of surgery another local midwife drives out to the house and sews her up right there on our own blankets. As I hold Anneliese's hand while the sutures are placed, I am grateful that we have been allowed this gentler transition. When the repair is complete, there is brief happy chatter. Then someone hands the baby back to Anneliese. Amy snuggles in between us, and we—we *four*—are left in quiet.

By dusk everyone has cleared out and left us alone in our old house. A local man who came here turkey hunting once told me his grandfather was born between these walls, and I try to imagine the birth scene then. No blue tub, I think, as I unspool the garden hose and siphon the water down the laundry room drain.

Amy helps me break the tub down and scrub it clean while Anneliese and the baby rest. Donna has made food for supper and several prepared meals for the days ahead, and in the fridge I see food containers left by my mom. Leah stayed to do several rounds of vital signs and assessments of Anneliese and the baby, and set us up with a bedside checklist of our own, including a sheet of paper listing every imaginable perinatal complication broken down in two categories: "Yellow Flags" and "Red Flags." I dared not read it, but I kept it close. Before she departed, Leah left a large jar of homemade bran muffin batter in the refrigerator. Donna baked a batch, and I thought it was a fine thing to be given the gift of a house filled with the smell of fresh baking.

Being in our own home on this, the first night of our child's life, is comforting, but without the official interruption of a hospital trip I am left with a formless sense of unreality—up the stairs we came without a baby, and now *looky here*. It's a soft-focus *Shazam!* Naturally, mingled with the glow in our hearts there is some trepidation, but at 10:45 p.m. the child poops. I take this as an affirmation of life.

At midnight, she poops again.

In the morning there is snow on the ground.

Leah and her apprentice return the following day to perform the newborn screen, and when they make the foot imprints to accompany the birth certificate, we get a taste of exactly what we have unleashed in this world. Unhappy with being dangled feetfirst in the air, the baby skips past crying and rockets straight to the furthest purple fringe of outrage. Such *blaring*. Not howling, not wailing, but a full-on sustained brass note fit to raise a regiment. Golly. It sounds like a blowout in the bugle factory. Today when Leah leaves, she reinforces something she has been telling us since we first met with her: Keep the week following the birth for yourself. Let the mother rest. No outside visitors. Not even well-wishers. It seems extreme, but we soon learn what precious advice it is. Donna stays to make meals the first couple of days, then she and Amy leave to visit relatives. Anneliese and I spend every day together. Nothing but us and the little one, for a week. We don't answer the phone. I stay out of the office and don't check e-mail. We learn the rhythms of the baby. Change diapers. Celebrate the glorious day of transition when the baby's poop changes from black to yellow.

It isn't a vacation by any stretch. There are some concerns early on—the baby is a shade jaundiced (Donna fixes that by sunning her in a chair beside the window), she has trouble with sucking (failing to establish, as I come to learn, a proper "latch"—what an apt application of the term!), and I am on the phone to Leah more than once this week with concerns about the comfort of both baby and mom.

There is also the matter of naming the child. We've been waffling for months. While Anneliese does her best to invest the decision with spirituality and ancestral reverence, I am largely concerned with scansion and assonance and the potential for naughty playground rhymes. Furthermore, it has always seemed to me that a child's name should be reducible to one crisp syllable for what I call the "freeze-factor," to be used when you wish to arrest the progress of the child in a precipitous manner, like when he is about to stick his fingers in the fan or she is sneaking out the bedroom window, in which case you want a name you can crack like a whip. *"Pollyanna!"* for instance, has no freeze factor. It got to be a bedtime game, the name list: Anneliese would read her latest choices, and one by one I would bat them down. Then she would do the same for me. There were some doozies, but I will not reveal the list of rejected monikers, because somewhere out there is someone else who dreams of naming a child Ezekiel Storm. *Zeke!* (I practiced.) On day five or six of our young child's life it becomes a matter of some embarrassment, and so we take the form the government provides, and—in honor of a family member—write "Jane." Then I try it out: *"Jane!"* The kid doesn't flinch.

Within the hour of Jane's birth, I snapped a photo of Amy holding her newborn sister. It wasn't posed or arranged, I just pushed the button. When I looked at it later, it took my breath away. Without realizing it, I had captured Amy just as she inclined her head to kiss her sister's brow. Her arms encircled the baby, her eyes were closed, and her lips were just brushing the crown of Jane's head. For her part, Jane is asleep in a nest of blankets, her chin resting on the curled knuckles of her left hand. I stare and stare at the photograph, my eyes wet. I am feeling blessed, blessed. But I think too of how so much of this world is the equivalent of busted concrete and twisted rebar, and I am jolted at what parents are charged with, and how limited our powers may be. Thankfully, Amy has a way of perforating my direst pretensions and lightening my worldview through the application of humor, intentional or not. Shortly after the beatific image was taken, she phoned her father Dan in Colorado, and fairly busting with pride, announced, "Well, you're a dad again!"

One lives in the glow of the miracle of new life and then rather harshly discovers that the electric bill is due again. We had our wonderful cocooned week, and even in the wake of that I was able to skirt deadlines and remain mostly home, but now real life presses back in. I have a raft of backlogged writing deadlines, volumes of unanswered e-mails, the usual stack of bills to pay, and I am returning to the road soon. I have always loved the road, and am still eager to feel the wheels beneath me, but nowadays my heart turns homeward sooner than in the past.

We plan to get the pigs when I return from this next round of travel, so I'm trying to finish up the pen. I've got it mostly enclosed with panels, but my brother Jed has recommended that I run a strand of electric fence all around the perimeter about six inches off the ground. The panels will hold the pigs fine, he says, but they are capable of generating great upward force with their snoot and shoulders, and if they get to rooting around the base, they'll boost the panels, posts and all. He grins when he tells me this, and you can pretty much picture him chasing pigs.

First I have to clear the way. I put most of the panels up when everything was still winter-dead. Now the nettles and burdock are knee-high. I don't own a scythe or a grass whip, so I have at them with a hoe, which is not pretty but gets the job done. I'm slashing away like a grass-stained Sweeney Todd when Amy ambles down. "Oooh, nettles!" she says. "Yum!" She watched Anneliese drink nettle tea throughout the pregnancy, and the two of them regularly collect nettles and bake them in our lasagna. This is all a reflection of our friend Lori the wild foods expert. Lori has taken her daughters and Amy on several foraging expeditions, and as a result Amy is forever eating dandelions straight from the yard or bringing me fistfuls of wood sorrel. The wood sorrel is evocative (as a kid I plucked it from a damp patch out where the sump pump drained) but a little too sour for my taste. The back of my hands and forearms are sweaty and tingling with nettle-sting, so it's nice to have Amy remind me of its happier attributes.

Once I've cleared away the foliage, I begin placing insulators. To save money on posts, I planned to secure the insulators directly to the panels, but first thing I discover is there is no way to do this without seriously modifying each insulator. I do a quick calculation of time and gas money versus the price of a bag of plastic insulators and decide to forge ahead. The required modifications involve profound misuse of a tree pruner, but it works (if necessity is the mother of invention, I am its ham-fisted stepchild), and before long I am placing the insulators while Amy follows along behind, happily hand-tightening each threaded retainer ring. During this time our old friend Mister Big Shot reappears, squawking and flapping around the perimeter. Amy rolls her eyes. I quietly hope she will learn to recognize similar chest-puffing inanity in the males of her own species and react with the same disdain. It's a long sail from six years old to safe harbor.

By the time we get all the wire strung and snug, it's nigh on suppertime and I decide I'll hook the power up another day. Returning the tools and fencing equipment to the shed, I see my beloved International pickup sitting over in the corner. The carburetor is leaking. I need to fix it. *Another day.* I notice the lawn

needs mowing. *Another day.* I'd like to fence off a big chunk of
the yard and get sheep. *Another day.* Through the screen, I can
hear Jane blaring.

We've been slowly emerging back into the world as a family. Relatives begin stopping by, and for the first time Anneliese's grandmother holds the baby. Grandma Scherer is ninety-four years
old and has only recently traded world travel for the Internet.
A preacher's wife who raised five children while holding down
a teaching job after her husband died young, Grandma is one of
those women who makes you feel sluggardly. When I leave the
room to get the camera, I return to find Grandma rocking Jane
and singing a lullaby in the original German.

Nearly once a day now someone will hold up Jane, look at me,
and say, "So—what do you think of the baby?" and what I want
to say and sometimes do is how above all the arrival of this tot
has only expanded the love I feel for my wife. The vision of her
pushing fiercely, then the sound that rose from her when first she
held that baby close—there is something of an eye-opening ear-
tweak in there for a man. I remember thinking, *lioness.*

Now, however, she is drawn and pale. After months of preg-
nancy-induced insomnia, she had been longing to sleep. And indeed, she has been able to sleep at night when the baby isn't
waking her, but during the day, during those times when she is
desperate for a catch-up nap, she simply can't doze off. Other
mothers are giving her plenty of advice, and at one point she says,
"If one more woman tells me to 'sleep when baby sleeps' . . ."

Sometimes to keep the house quiet during the day when Anneliese is trying yet again to sleep I strap Jane into a red quilted

baby sling Anneliese's mother used to hold her babies. I checked
the label and it was made in the 1970s. It has clunky stainless
steel clips. But it works great, and I am able to write for long
stretches with the baby asleep against my chest. Recent research
has cast some doubt on the benefits of playing classical music
for unconscious infants, but I have my own ideas, and today
while she snoozes, we are edifying ourselves with a rotating mix
of Dwight Yoakam, Clarence "Gatemouth" Brown, Mark Ches-
nutt, Greg Brown, Loretta Lynn (for spirit), and Iris DeMent
(for unvarnished holiness). And in the interest of imbuing more
ineffable feminist sensibilities, I pulled Cinderella's *Long Cold
Winter* and replaced it with Shawn Colvin's *A Few Small Repairs*. Jane sleeps peacefully, rousing only to move her lips and
make a noise somewhere between snoring and drooling best described as snurgling.

Not so long ago I stepped through the front door to find Amy in
the middle of the kitchen unrolling a flag-sized poster of me. It
was from a book tour stop somewhere back along the line. My
visage was full-color and big as a cheese platter. Amy held the
poster unfurled before her, and I admit I savored the moment
right up until she turned and laid it faceup on the bottom of the
guinea pig cage. I am well aware that on a scale of one to Britney,
I peg the fame meter roughly three notches below the lieuten-
ant governor of Maine, but even so this was a severe calibration.
"WHAAAAT?!?!" I said, theatrically feigning great dismay. Amy
giggled and scattered wood chips over my gap-toothed mug.

Now it's another cage-cleaning day. When Amy finishes, I help
her place the guinea pig back inside with his bowls and purple

plastic igloo. After securing the lid we return the cage to its cus-
tomary spot in a corner of the living room and go outside to check
the progress of the seeds we planted in the cold frame. The radishes and lettuce came up two days ago, and today we find the
spinach sprouted. We're in a run of cool breezy days but the sun
is out and condensation has formed on the glass, so I prop it open
to let it breathe. Amy and I meander across the yard and down a
slight slope to a lane formed between a dense row of spruce trees
and the south-side wall of the pole barn. The spruce block the
breeze, allowing the steel to gather heat from the sun. We press
our shoulder blades against flat spots between the vertical corrugations and slide down to sit and soak up the warmth. The buffer
zone of spruce muffles the rest of the world. "There was a place
like this out behind Grandpa's barn," I tell Amy, thinking of a
nook between the silos where I loved to hunker as a child. There
were weeds and a patch of sand. I liked to sift the sand through
my fingers, the flowing tan grains speckled with bits of bright
green shingle grit dislodged from the barn roof by generations of
rain. I tell Amy I sought the silo spot on early spring or late fall
days when you need a windbreak if you want to feel the sun.

"This could be your place like that," I say.

"Yes," she says, plucking at a weed stem. Even as she answers,
I know I'm pushing a rope. She'll have to find her own places.
I mustn't assign memories. We sit and visit, and as invariably
happens I find myself stowing the moment for all the road time
to come. The scaredy-cat part of me wonders if she will do the
same. And if so, will the memory warm her or simply sharpen
my absence? When we walk back up to the yard, there is a blue-
bird in the maple tree beside the corncrib. I point it out to Amy

and she locates it easily. When pursuing the heat of the sun I must never forget it exists also to illuminate blue birds in brown branches.

Later in the afternoon I find the glass lid of the cold frame smashed. I suspect Fritz the Dog. He was nosing around earlier. Fortunately I have a fair collection of old storm windows, so I gather the broken glass, install a replacement and prop it open again. When I see him lurking in the same spot again later, this time with a chewy dog treat in his jaws, I holler at him and shoo him away. But when the day cools and I go to lower the lid, all the dirt and most of the seedlings have been scraped into a mound in one corner. I realize now he's been looking for a soft patch of dirt to bury his treasures—I'll lay odds there's a dog treat under that mound of dirt. The dog is nowhere to be found, so I can do him no harm, but I am ashamed to say I storm into the house and slam the door and say something very loud and forbidden. I can't defend my rage, but it is tied to the fact that in the midst of all that has been going on, and all my absences, that little plot of dirt with its sprouts was a tangible manifestation of some careful moments spent with Amy. I don't care about the stupid plants, but I care about what it meant to kneel down there with my daughter. Later, when I have cooled down some, I go back out and notice the dog missed about six radish sprouts. I lower the lid and figure maybe they've got a shot. Then I go for a cool-down walk. Along the south side of the granary, the rhubarb is up. The last time our family gathered, my brother John—a big bearded fellow who spends a lot of time on a bulldozer—said he eats an entire rhubarb stalk every spring just for the involuntary face-scrunch that transports him back to his preschool days. He also reminded me that having heard rhubarb leaves were poisonous, we would feed them to the chickens and then hang around to see what happened, but nothing ever did.

The baby has cried us awake. Fumbling in the dark to fetch her, I note the eastern horizon is a faint charcoal gray.

Early to bed, early to rise has never been my deal. Half of everything I've ever written was likely typed past midnight. Not so any longer. Age plays a part, but mostly I think it is a sequela of parenthood. Even before the baby, when it was just Amy, I had begun easing toward the early shift. Writing after supper, I'd take a break to read books with her at bedtime, and find it near impossible to go from that quiet moment back to the desk.

The new sleep pattern has been reinforced by the baby crying at night. After twenty years of going from slumber to blastoff at the first micro-beep of an ambulance or fire pager, I tend to spasm straight up and out of bed at Jane's least whimper. Anneliese is bemused at the gymnastics, which is to say that while she appreciates my willingness to help (it's less about helpfulness than doggish conditioning) she could do with more *arising* and less blastoff. Furthermore, in most cases the baby is looking for the drink I cannot provide, so although I wake to retrieve her, by the time she is nursing I've returned to unconsciousness. She howled at 2:00 a.m. and now she's howling again. I consider the dim seep of light and decide I might as well begin the day. By 10:00 a.m. I'll be nodding off above the coffee cup, but for now I want to get going.

In soft lamplight I place Jane at her mother's breast and lean down to kiss them both on the brow. Jane's cheeks are fattening, and when her eyes open I look for recognition but I still don't quite see the person in there. I wonder if it's just me or if mothers attach from the first instant while the man flounders around and waits for the fun stuff, like diaper farts and jibber-jabber. I poke my head in Amy's room and in the glow of her night-light see her wrapped in a sleeping bag on the floor beside her made bed. She has taken to doing this since the baby came. Still impaired by a developmental psych class I was required to take in college, I momentarily worry that the change may be portentous; then I decide it's possible the kid just wants to sleep on the floor.

Downstairs, and out the door. Eastward the gray band is lightening, but the sun remains well sunk. Drawing the cool breath of morning into my lungs I think of my father, whom I do not believe has missed a sunrise in some forty years and would be startled to find me up and about at this hour. I still love the dark heart of night when it is possible to believe you have the world to yourself, but I can understand why Dad loves to watch the day come in. And I find I am a little less breathless working from this end of the cycle than I am trying to fight my way through to some sort of bleary-eyed finish at 3:00 a.m. There is the idea that you have a head start.

When I get to my desk I power up the computer and open my e-mail. As the new messages roll in, a simple subject line catches my eye: "Tim."

The e-mail is from the sister-in-law of a dear friend in England. I double-click it.

Hi Mike,

Some time ago Tim was diagnosed with cancer of the liver and was told that he hadn't got long to live. He chose not to tell you as he wanted you to remember him as he was.

Tim passed away on 20th April at 3.am, he died as he wanted to without any fuss.

We weren't sure how to contact you as you are often on the road and thought this was possibly the best way.

Don't know what else to add at this point, we are sorry we know this will come as a shock Mike, but I know we will talk very soon.

Claire Amy Sylvia and Ronnie.

Aw, *Tim*, I think. I raise my eyes to the wall directly across from the desk: Tim, in an old photograph framed and hung from a nail. Twenty-three years we were friends. Last time I saw him he was fine. I check the date in the e-mail again. Six hours' time difference—he would have died last night while I was frittering at the end of day.

When my mother was a child, she had a passel of international pen pals. Over the years the correspondence waned, but she and an English girl named Pat kept in touch into adulthood. In 1984, fresh off my first year of college, I traveled to England and my first stop was at Pat's house. Pat had two daughters. One of them was dating Tim. We met the night I arrived, went to the local pub together the following evening, and got on like well-worn pals from that time forward.

His given Christian name was Timothy Swift. I always thought this an eminently toff English moniker, but you wouldn't peg him to it if you saw him in the pub. There was nothing Jeevesy about the boy. He was a resident of Cannock, England, a Midlands lad, born near enough the environs of Birmingham that he carried the working-class Brummie accent (think Ozzie Osbourne with a cold), although how much of his accent was geographical cottonmouth and how much was just Tim is hard to know. Even his friends and relatives frequently found him indecipherably mumbly. I spent enough time in his company over the years that I grew to understand him relatively well, and during his visits to the States I happily served as translator. My advantage lay in the fact that the night we first met, Tim was convalescing from having his four upper front teeth knocked out in a pub parking lot the night previous. From my perspective, his locution only improved thereafter.

We called him Swiftie. He stood maybe five-four, favored Motorhead T-shirts and black socks with his tennie trainers, and wore a rose tattoo on his forearm. The rose was smudgy and prone to bubbling in the sun. The year we met he had just completed the English equivalent of technical college and was working at a factory, building motorcycle frames. This was a great relief to his mother, as a few short years previous he had been a greasy-haired headbanger with no evident prospects of a legal or supportable sort. In the one photograph I ever saw of him from that earlier era, he was devil-eyed and grinning around a remarkably misaligned cluster of incisors. In fact, he once confided that although he might have preferred a more professional procedure, having his teeth head-butted to the tarmac was actually a bit of a windfall, as the court instructed the other fellow to purchase Tim a new set that in the end were implants of model quality.

At the desk, still staring at the e-mail, I'm going back, in filmstrips and flashes: Tim and I walking home in the dark after the pubs closed, stopping at the bright-lit chippie off Longford Road. Undoing the tight-wrapped packet and eating the sodden fish and potatoes straight from the paper while watching *The Young Ones* in a room smelling of hot grease and vinegar. In 1989 we wore garbage bags and stood in the rain for hours before finagling our way into Centre Court of Wimbledon under creative pretenses. One moment we were sodden proles, the next we were seated within full view of a duchess. Tim got the better seat, but sadly he was spotted and bounced almost immediately. As the guards escorted him past me, we studiously avoided eye contact as previously agreed and I subsequently enjoyed the entire match. Edberg versus Mayotte, if my memory brackets are accurate. On the way home from Wimbledon well after midnight, Tim's car broke down on the motorway. A late-arriving tow truck took us deep into the countryside and pulled inside a barn, at which point the furtive mechanic pinpointed a problem with the clutch and named a ransom for repair. With the same easy mumble he would use to request his fifth lager, Tim told the guy to bugger off and drove us home clutchless, his trucker-shifting not impeccable but serviceable, and hours later we lurched through the final traffic circle and herky-jerked to a stop in the driveway at dawn. Another of my visits coincided with the rise of electronic trivia games in the pubs, and our combination of wit—Tim's in science, engineering, English sport, and culture and mine in fluffy minutiae—did not make us rich, but did regularly enable us to pay for lunch. How solid the pound coins sounded when the machine chugged them into the tray. We road-tripped to Wales and the Lake District, hiking for miles in the rain, sleeping in a damp tent, and stopping to eat in pubs where the patrons switched to Welsh upon our entry. Tim was a serial hobbyist—one year darts, another year winemaking, next year the curry club—with a tendency to immerse himself headlong (learn all the lingo, get all the gear) before abruptly moving on. During his competitive fishing phase I accompanied him to a canal-side tourney where he diddled at the water with an absurdly long pole and used a slingshot to launch maggot clusters across the channel as chum. Another time he joined a sporting clays club and took me on a round, reveling in our rare good shots by adopting the Queen's English: "Jolly hockeysticks! Bag another grouse, Jeeves!" Oddly enough, when mimicking the Queen, Tim was quite understandable.

One very late night after everyone had been drinking with the exception of square, teetotaling me, I chauffeured Tim's girlfriend home while Tim trailed behind on a moped he had resurrected from a junk heap. (Whether it is more dangerous to allow a sober-but-right-lane-imprinted Wisconsin rube to navigate the narrow roads and traffic circles of suburban Britain while attempting to shift a dodgy left-handed manual transmission with his nondominant hand at 2:00 a.m. or to yield the wheel to a tipsy native is a conundrum to be parsed another time—we were young and predictably senseless.) Tim had got through some lager, so I kept checking his one wobbly headlight in the rearview mirror. A kilometer from home, I looked up, and the light had disappeared. We circled back. Shortly our own headlights illuminated Tim, placidly pushing the moped along the dark street. As we drew nearer, I could see the bike was bent and badly scratched.

"What happened?" I asked.

Tim looked at me, a little blurry, but wholly unperturbed.

"I f'got t'balance," he slurred. And then he pushed off into the night.

We circled again, caught up, drove slow beside him, and saw

him safely into the house. When we left he was staring into the open fridge, contemplating a sandwich.

The sun is fully up and bright. It is early afternoon in Cannock by now, so I call Tim's mother Sylvia. It was very hard, she says. He wouldn't let us contact you because he knew it was going to be bad, and it was. He suffered terribly, she says. Sylvia and I talk a little longer. Then I hang up and try to work. Anneliese's mother is visiting, and has made breakfast. As she often does when I work mornings, Anneliese comes to my office with a plate. I thank her, and take the food. She looks at me and senses something.

"You OK?"

"I got some bad news . . . ," I say, and then the choke in my throat turns to tears. When Anneliese and I were married, Swiftie made the trip. Flew transatlantic cattle rate just to land on Thursday and leave on Sunday. The day he arrived, we spent the night in a tiny shack in the middle of forty acres near my beloved New Auburn. The next day we copiloted my old International pickup down here to Fall Creek to prepare for the wedding. The morning of the outdoor ceremony Swiftie helped my father-in-law Grant and me set up the chairs and then take them all down and reset them in the tent when the weather turned to rain. After we finished I headed to the house for a shower, and, looking back, I saw Tim at the edge of the lawn beneath the tent, smoking a hand-rolled cigarette and looking out across the sweep of the valley below. His free hand was in his pocket, and he was rocking one knee, the way he always did when he was relaxed and taking something in. How many times I had seen him in that stance, raising and dragging at the cigarette without hurry. He'd keep the knee going and bend at the waist a little, like he was working up a bow, but then in the end he'd just muster a faint smile, his lower lip slightly pouted, his eyes squinting as he raised the cigarette again. This morning when I read that e-mail, the first image that flashed—even before I looked to the photo on the wall—was of Tim on the hill there, quiet, alone, content.

I wonder if he knew.

Experts say the honeybees are disappearing, so it's nice to see them busy at the bush beside my office door in the early afternoon. I cannot identify the bush—it verges on shrubbery—but on this the day of my friend's death it is in bloom, the modest yellow blossoms waxy in the sun. Noon has passed, lending the light just enough postmeridian slant so when the bees buzz by, their minuscule shadows trace across the window screen like silhouette radar. It's a gentle sight, enhancing the sun and easy breeze. The bad news from England has had the immediate effect of compressing the world and time. I've kept at my work, but am continually drawn down memory's kaleidoscope wormhole. Feeling the need to walk in open spaces, I leave the desk and head for the ridge.

Sylvia said Tim came back to his boyhood bedroom to die. I know the room. I can go there in my head. I bunked in the bed there sometimes. I suppose he did it to spare his young daughter Amy and wife Claire. I don't know. His Amy was a toddler last I saw her. I'm walking and walking, farther and farther back on the property, into a valley not visible from the house. The air is warm. Deep in the trees, the air smells of duff and thaw. I wish he had called me.

He wanted you to remember him as he was, it said in the e-mail. When I spoke with Sylvia this morning, her words were exactly the same: *He wanted you to remember him as he was.* I think of him in the yard with that cigarette and how much I could read from just the jiggle in his knee, and yet our span of two decades was built on less than a hundred days spent in common company: there are implicit questions of depth. By the end he had become successful in his field, managing international projects for one of the largest engineering firms in the world, but only once did I see him at work; I was caught off guard by the man in the tie and white hard hat. He oversaw a tunneling project beneath the English Channel, and ramrodded another in which slurry was pumped at extremely high pressure into miles and miles of abandoned underground coal mines. Once the pipeline blew and took off a man's arm. Tim hit the kill switch and grabbed the arm. Another time he got a frantic call from the manager of a high-end car dealership screaming that his slurry was blasting through a hole in the middle of the showroom floor. Tim loved telling that one, but eventually he was promoted to the point where his job amounted to serving as shock absorber between middle management and the uppermost tiers, and it wore on him. The better the pay, he said, the worse the pressure. He spent most of his days on the phone, translating vituperation. The last time we talked, he said he was going to give it up. He talked about his Amy, and Claire, and how too often the work kept him away from home.

At the far end of the valley I begin the looping climb back, topping out in a patch of popple. And now I'm crying. I wish he had let me know. I wish I could have seen him one more time.

The usual selfishness of grief. I am not angry, I am yearning. Overhead the tree trunks fork their dusty white bark into sunlit greenery, the newborn leaves limp and luminescent somewhere just short of chartreuse. A shifting scatter of light plays across my head and shoulders, and I am grateful for the cathedral feel of this place. Grateful that I might grieve in natural sanctuary.

I have a good sweet weep. Then I walk back to the house. I want to hold Jane. Feel life in my arms.

A few days later I am on my way out the door to hook up the electric fencer. Anneliese is on the couch with Amy. They are reading *Beetle McGrady Eats Bugs!* "Stink bugs taste like apples!" says Amy. "I'll take your word for it," I say.

I've mounted the fencer on a post inside the pole barn, to keep it out of the rain. The power unit is an unremarkable plastic cube the size of a half-pint ice cream box. When I plug it in, a pinpoint green light glows on and off, indicating that the fence circuit is complete. The fencers of my childhood were more the size of a twelve-pack, and were commonly housed in stylized tin shrouds. One resembled the front fender of a Ford Fairlane. Another of my favorites was dusty blue with a silver-riveted logo plate and a fat orange indicator light that eased languorously from lit to unlit. I used to stand in the barn at dusk staring up at the deliberate amber blink and imagine the unit was an advance robot broadcasting homing pulses to the distant mother ship. Dad's first fencer was called a Weedburner, an apt name considering that shortly after he plugged it in, flames swept the pasture and there were fire trucks in the back forty. In a nod to my father's frugality, years later I would be out fencing and find myself threading the wire through partially melted insulators remnant of the fire.

Unlike many a curious farm boy, I swear I have never peed on an electric fence. I am told this blows a very specific fuse. Perhaps the act prevents prostate cancer—a longitudinal study is in order, challenge number one being the location of subjects willing to 'fess up. I do remember walking down the barnyard lane with a steel can of *Off!* and trying to see how close I could run it past the fence without making contact, a diversion that lost its appeal when I got knocked to my knees. To test the steadiness of his hand, my brother Jed once formed a circle around the wire with his fingers and took off walking only to zap himself flat, establishing that neither his intellect nor his fine motor skills would qualify him as a brain surgeon. One of the Carlson boys used to check to see if the fence was energized by slapping at it with his open palms. He swore that if you touched it quickly enough the shock was minimal. We assigned him special powers until the day he mistimed his swat and his hands clenched around the wire in an electrified spasm. The current would break just long enough for him to begin unwrapping his fingers, then the "on" cycle would hit and his hands would seize into fists again. Hearing the howls, his father ran to detach him.

On one of my prior Farm & Fleet runs, I purchased an electric fence tester consisting of a slim grounding wand connected by a coated wire to a plastic paddle tipped with a copper terminal. You stick the wand in the dirt and touch the terminal to the wire. There are four indicator lights mounted in the paddle; the more zap your fence generates, the more lights are illuminated.

Sadly, despite the fact that I quite uncharacteristically read and reviewed the written instructions during both the installation of the fencer and the wiring of the three ground posts, I can only get two of four lights to illuminate. I recheck the wire from fencer to the farthest termination point—all clear. I recheck the ground posts—everything is in order. Still only two lights. I have no idea if that's hog-worthy.

In the end, I test it the way the old-timers taught me. Plucking a leaf of green quack grass, I grip it between my thumb and forefinger way back at the stem end and lay the pointy end across the wire. Then I slowly push the green blade forward until I feel the first faint tingle. What you've got here is an organic rheostat. As the quack blade advances, the resistance decreases, and you get a better zap. How far you keep pushing is up to you. When my knuckles are about four inches from the wire it feels like someone is snapping me in the wrist with a rubber band. I figure that'll hold pigs.

My plan had been to get back at the work waiting in the office, but I dive straight into fencing the garden. The rabbit population around here has been exploding. With no barrier they'll decimate our vegetables. And the planting season is nearly upon us. At least these are the things I am telling myself. There is some truth to it, but there is an unquestionable element of escapism. When I get way behind on deadlines and responsibilities as I am now, I rather perversely throw myself into physical labor, which yields palliative sweat and tangible progress even as I fall farther behind.

While shuffling through a pile of mail and miscellaneous papers beside the telephone today, I came across some scribbled notes. They were in Anneliese's hand, and appeared to be the rough draft for a set of talking points: *tired baby/tired mom/7-year-old = frustrated mom . . . things you can do . . .* The subsequent notes essentially sketched out something that Anneliese brought up recently, saying she appreciates everything I am doing to pay the rent and prepare for having animals, but sometimes she wonders if I'm using work as a hideout. It made me crabby that she would even suggest such a thing, because of course it is true. I can provide plenty of justification—a man must Provide, soon I'll be on the road again, yada yada—but there is no question I find refuge in the work, and I'm not sure I've got it in me to change on that front. I love to put my head down and bull.

Amy appears, apparently still ruminating on the bugs of Beetle McGrady. "Mommy says when she was in Mexico she ate a taco with crickets!" She is bursting with wonder and admiration. I wonder how often Anneliese wishes she were back in Mexico or even her Talmadge Street house and not saddled with an irritable self-employed scribbler wiring slapdash pigpens up a dead-end road.

Plunging into the garden plot, I rip up last year's weeds and clear the overgrowth from the perimeter. Then, using a posthole digger, a level, and a two-by-two tamper, I set the poles (salvaged from where I found them leaning in a corner of the pole barn) solid and square in the dirt. Next I dig a trench so I can bury the bottom few inches of the fencing to prevent the rabbits from tunneling under. After stretching and stapling the fencing in place, I fill the trench and stomp the dirt down flat all around. Lastly I rig a gate. I keep my head down, working steadily, sweating and not stopping. There is far more in play here than work ethic. A teacher of psychology once reviewed my behavior over the long term and pegged me for bipolar—it strikes me that this desire to hide out by hooking oneself to the plow may be nothing more than the manifestation of mania. I once knew a woman whose manic swings drove her to don scarlet clothes and makeup and dance the downtown streets, whereas your manic Scandinavian will dig postholes.

With the garden enclosed, I bring out the rototiller. For the next half hour I wrassle it back and forth until the patch is fluffed and soft and ready for seeds. Anneliese has been reading up on reduced tillage, mulching, and cover crops, and we intend to move that way, but for now the plowboy in me is soothed by the pillowy look of the churned earth. I step back for a moment to take it in and am heartened by the solid set of the posts and the taut lines of the well-stapled wire. I am forever cobbling things together—it feels nice to look at a job and think it might last. Amy has stripped down to her underpants, lain flat out in the fresh-tilled soil, and is sweeping handfuls of dirt over her legs and tummy. I start to tell her no, then walk away and leave her to it, one of the better decisions I've made all day.

I store the rototiller and walk into the house, dirty, sweaty, thirsty, hungry, and surprised to find several hours have passed. Upstairs I can hear Anneliese pacing and Jane crying. I wash up and take the baby. Laying her belly-down along the length of my forearm, I grip her torso with my hand. We call this the football hold, and it is the one thing I seem to be able to do well, babywise. Her arms and legs dangle awkwardly, but she nearly always settles and quietens, and does so now. Perhaps it is simple syncope. Soon she is asleep.

My mother-in-law and sister-and-law are in the kitchen making venison stir-fry. When it's ready we eat on the deck overlooking the valley. Jane is awake again and happily gurgling. We're letting her air her little hinder out, and she celebrates her diaperless freedom by peeing on the tablecloth. A minor diversion compared to this morning, when I was washing her on the changing table and with neither wink nor warning she ejected a rope of poop that arced into the wall six feet away. A true hydraulic marvel.

After the hot, sticky afternoon, storms have begun working either side of the valley and pushing a cool breeze before them. It's nice, all of us out here together, eating and talking, laughing with the baby. I get going on the pigpen, or the garden fence, and from some imaginary omniscient perch I look down and see a man toiling on behalf of his family, forgetting that sometimes what the family needs is a man sitting still.

In the summer of 1989 I lodged with Tim's parents for a stretch. I was trying to become a writer at the time, and began every morning in the front room, drinking tea beside a glowing coal grate and clacking away on a manual typewriter lent me by Tim's mom. Tim had only recently moved out, and his turntable and a collection of vinyl albums remained on a low shelf beneath the windows that opened out to the street and front garden. Slice by vinyl slice, I worked my way through the music. Last night while writing under deadline, serving the clock more than the muse, I procrastinated by going online to track down a copy of Marillion's *Misplaced Childhood*, an album I hadn't heard since those mornings on Longford Road. The tears came at the chorus of "Lavender" (*"Lavender's blue, dilly dilly, lavender's green . . ."*), but they were tinctured with gratitude that a song might so wholly transport me back to my friend.

And so this morning I spent an hour in the pole barn digging through the boxes where my music CDs have been stored since the move. Box by box I flip through the jewel cases, scanning the spines and pulling anything that evokes our long-gone days: the Waterboys, from my first visit in 1984. Simple Minds, for whom the Waterboys were opening the drizzly day we saw them in Milton Keynes Bowl. *Avalon*, by Roxy Music. Pink Floyd's *Animals* and *The Final Cut* (Tim put me onto these after finding me listening to *Dark Side of the Moon* for the sixth day in a row). Siouxsie and the Banshees. The Cure (thirty seconds into "Plainsong" and I am alone in the Longford Road front room at 3:00 a.m., staring out the window at yellow lamplight reflected on wet tarmac, the rain gone to mist). I pull a Bronski Beat album so I can revive the wash of summer traffic and the scent of daffodils weaving through the second-story window of my British bedroom, matched forever with Jimmy Somerville singing "Smalltown Boy," his voice plasticky on the clock radio beside the bed as I wrote in a notebook and listened for the sound of Tim's car pulling in the drive.

By the time I head back up to the office, the stack is such that I must steady it with my chin.

It's a fine line that separates wallowing from remembrance, but as I listen to the music for the rest of the day and into the night, I don't care. Track by track, I am back with Tim, riding shotgun in the left-hand passenger seat, strap-hanging on the Tube through London, or simply scuffing home from the local. It's a mind trick, and I'll take it. When the three-chord stomp of "Rollin' Home" comes thumping from the speakers, we are together again, driving back from a Status Quo show at the National Exhibition Centre, smiling young men on the road to who knows.

He wanted you to remember him as he was. A clichéd phrase intended only for my comfort, I thought when I read it in the e-mail and heard it on the phone. But the music is working on me, and I'm beginning to understand. I'm thinking of Tim, dying in that room I knew so well. How in all our years of coming and going, we made it a point to never say good-bye. He knew if he called me about the cancer I would want to come over. If I had come right away, there might have been time for a few more nights, or a few more miles, but it would all be building to the inevitable stilted good-bye, my very presence reminding him he was bound to die. If I had gone over in the later stages, he would have been too far gone, in too much pain, and I would have done him little good. If I was unprepared for his death, I am just now realizing he wanted it that way. Tim fooled me. Fooled me beautifully, and gently. *He wanted you to remember him as he was*, and I will, and do, because he gave me no other choice. He did not choose his death, but he chose his exit.

In a feeble effort to hold up my end of the domestic partnership, I am doing the dishes after supper and notice Amy wandering out the driveway with the dogs. In ten minutes she returns. Standing with the door open, she says, "Do ants have protein?"

"Yes, they do," I say, turning from the sink to look at her. "Why?"

"Because I ate one."

"Really!"

"It tasted sour."

So she chewed it, then.

The days have cycled through. The maple buds have unbundled. The hills are a green divan buttoned with clusters of bloom that foam up apple-tree pink and chokecherry white. After lunch I am trying to allow Anneliese a nap. She is upstairs, and I am downstairs with Jane across my lap. The deadlines have stacked up, so I am also trying to write, the computer balanced on my knees. But of course I can do little more than study the baby. Her sleepy tics, her bursts of rapid eye movement, her bow-perfect lips, her candy-floss hair glinting auburn in the sun. Her nose is resting on her knuckles, and her head rocks slightly with each breath drawn. I am playing music on the laptop: Innocence Mission. The volume is way down. The song sounds tinny and faint. I am studying Jane's impossible ear—this perfect miniature conch, a leaf just partially unfurled—when the final chorus repeats, barely audible: *"this is the brotherhood of man . . . this is the brotherhood of man . . . this is the brotherhood of man . . ."* When I turn my eyes to the valley below the big window, it is beautiful for a moment and then all the blooms and green dissolve in a watercolor wash.

After a suitable interval, the guinea pig whistles and flips his purple plastic igloo.

CHAPTER 6

Today a dog bit me grievously upon the ass. I apologize for the salty talk, but it was a galvanic moment.

I was wrestling a pig at the time.

So—two firsts in one day.

I have had my heart set on owning pigs for a while now, but as with so many of my projects, reality has taken a backseat to cogitation. A lovely thing, to sit back and ponder what One Shall Accomplish without having to actually lace one's boots. To price sausage makers prior to carrying a single bag of feed.

Farmers though we are, my family is short on pig experience. Dad didn't care for the smell of them, so we never raised any. Most of the farmers around here used to keep a sow or two, but they were being slowly phased out in favor of cows and crops by the time we came on the scene. I have a fragmentary memory of peeking into a penful of piglets over at the Norris North place when I was a toddler. Dad may have lifted me to look over the barrier, because I retain an omniscient perspective of the litter startling below me, shunting away in a single flowing motion, like a school of frantic pink minnows.

My sister Kathleen and her husband Mark have raised a couple of pigs each of the last few years. Mark does the butchering himself, using his skid-steer bucket to hoist the carcasses for the evisceration and skinning. My brother John hasn't raised pigs for a while, but he often tells the story of the first pair he butchered. They were brothers from the same litter and had shared the pen every day of their lives. On butchering day John shot the first pig between the eyes, and by the time it hit the ground, the other pig was licking up the blood and nibbling at the bullet hole. John used to raise his eyebrows and reenact the scene as if he were the bemused survivor: "Huh! Fred died!"

Then, half a beat later and still in character, he'd brighten: "Let's eat 'im!"

My youngest brother Jed raised pigs for several years but recently sold them all with the exception of his favorite sow, Big Mama. The sow hasn't farrowed for a while, but he keeps her on pension because he doesn't have the heart to ship her. Big Mama is approximately the circumference of a backyard LP tank and nearly as long. These days she is docile and grunty, but when she was younger, Big Mama had a litter and began to savage them. Our family grew up reading the wonderful James Herriot veterinarian books, and someone recalled a story from *All Things Wise and Wonderful* in which an old farmer dealt with this very problem by procuring a bucket of ale from the nearest pub and letting the pig drink herself docile.

Having neither pub nor beer at hand, the boys sent Dad to

town for a twelve-pack. This was great fun in light of Dad's tee-totaling ways, which had become quite public a few years back when one of the incumbent town board members encouraged him to take his turn as a public servant and Dad agreed, but first vowed he would never sign a liquor license. No problem, said the official, only two of the three people on the board are required to sign. Dad was elected to the board, and sure enough shortly thereafter one of the other two board members needed a liquor license. Ordinances stipulate that you can't sign your own liquor license, so Dad was on the spot. He didn't sign. So you can appreciate the murmur when New Auburn's one-man temperance union approached the counter at the Gas-N-Go and plonked down a case of Ol' Mil. I imagine the word reverberating up and down the street: *"Seriously! Bob Perry! Swear t'God! Hittin' the barley pop!"*

But the sow guzzled the beer, and it did the trick. With half a jag on, she let the piglets nurse in peace. Everyone was happy: the sow got a snootful, the piglets got dinner, and my brothers got themselves a good story to tell. As did the garbageman, when he dumped Bob Perry's recycling bin the following week and noted a smattering of crushed beer cans.

In all the buildup to getting pigs I have pretty much exhausted the reserves of my brothers and brother-in-law, peppering each with question after question regarding the housing, care, and feeding of porkers. Fortunately they are men of patience who furthermore have learned over the long term that their indulgence will be amply repaid by the quality of entertainment provided by my incompetence once I get rolling. These are men who can build things and fix things. I am convinced they frequently convene outside my presence to compare notes and shake their heads in wonder. They have so far stopped short of poking me with sticks.

So I have been talking pigs for months. But now the time has come. I had visions of myself trundling the pigs home in the back of Irma, my 1951 International pickup, but I still haven't fixed the carburetor, so I will take my mother-in-law's Chevy. Amy clambers happily into the truck beside me.

The last time I visited Equity Cooperative Livestock Sales, I was the same age as Amy. Dad didn't come here too often. He usually shipped cows and calves with a hauler, and when he had lambs to sell he drove them to the stockyards in St. Paul himself. (Sometimes I got to make the trip. I remember sitting beside Dad as the truck labored west on I-94 and he taught me to identify the make of the oncoming big rigs by the shape of their hood ornaments. We'd keep a running tally in his pocket spiral notebook. It's a game I still play and have taught Amy, but consolidation has taken most of the fun out of it—whither Autocar . . . Marmon . . . Diamond Reo?) But when it came to cull ewes, they weren't worth the shipping cost or the mileage to Minnesota, so he'd bring them to Equity. Plus, he told me recently, the sale barn provided a day of cheap entertainment for us kids. Like the zoo, with no admission fee or cotton candy vendors.

Today the rigs—mostly dusty four-wheel-drive pickups hooked to aluminum goosenecks—are of a different vintage, but they clog the parking lot in the same arrangement I recall from thirty years ago. When Amy and I step out of the pickup the gravel is white in the sun. All the empty trucks and trailers lend the lot a detached stillness, implying as they do that all the action is inside, out of sight.

I have arranged to meet a man named Kenneth Smote. Kenneth's last name always conjures some past-tense act of God. In fact, Kenneth is an atheist goat farmer and retired former chair of the local university psychology department, and father of my dear friend Frank. Over the years Kenneth has bought and sold goats at the sale barn, so I am hoping he can guide me through the process. Between critters, I envision an energetic discussion of fixed action patterns, specifically as they relate to the principles of imprinting as proposed by Konrad Lorenz—even more to the point, what are the odds that any given feeder pig will develop a lasting attachment to my favorite rubber barn boots? While we wait outside for Kenneth, I tell Amy that the sale barn used to be located well out into the countryside. The barn itself has not moved, but now it is within hollering distance of a mall. To the unexpected wrinkles of existence add the fact that slaughter hogs are available three minutes from Victoria's Secret.

Kenneth arrives in a worn gray Nissan sedan. An erudite man of comprehensive intellect known to write pleasantly eviscerative letters to the editor of the local paper, Mr. Smote nonetheless cuts an unprepossessing figure and comports himself likewise. He presents himself this morning in green coveralls, a cockeyed St. Louis Cardinals ball cap, and a wispy beard. After a pleasant hello and introductions—he and Amy have not met previously— we walk through the glass double doors of the foyer and up the steps to the sale ring.

Dad was right about the sale barn as entertainment. The

minute I hit the steps and smell the manure and sawdust, my pulse quickens. The seats are stair-stepped around three sides of the ring nearly to the ceiling. The front row seats are cushioned and fold down just like in a movie theater. The auctioneer sits ensconced in a stagelike enclosure with a microphone propped before him. There are cows in the ring when we enter, and we watch for a while to get a sense of the rhythm of the sale and figure out the bidding. Each cow comes in through a gate on the left, takes a few turns around the dirt while the auctioneer recites salient details, and then the bidding begins. The tension and gaming of the bidding charges the room. The rattle and rhythm of the auctioneer creates a breathless momentum, and now and then over the more organic scents we catch the smell of hot dogs and onions sold at the café downstairs. The bidding culminates, the winning bidder's number is recorded, the cow exits stage left, another enters stage right, and the drama begins anew. I've been to a fair amount of farm and household auctions in my day, but this was different. I couldn't keep track of the bidding, or grasp the process. We watch them sell cows for a long time. Then I take Amy out on the catwalk.

The catwalk is accessed through a door situated on the upper grandstand level. When you pass through the door and step out on the expanded steel grate you are essentially backstage, overlooking a vast holding area. Leaning over the pipe railing, we can see cows, calves, sheep, and goats. We look for pigs but don't see them. Finally, when we have traversed nearly to the end of the elevated walkway, we spot a pair of gigantic mama pigs, and a single litter of teensy ginger piglets. Trouble is, I'm looking for feeder pigs. They run about forty pounds. These big pigs are *too* big, and the piglets are too small. And I have no idea what that size of pig is worth. Or how I bid for just one or two. I don't want to wind up with the whole batch. I did get a bidding number before I came in, but I realize I don't even know what to do if I win the bid, and Kenneth says it's been a while, so he's not sure either. Then I'm making my way back up the catwalk when I come nearly face-to-face with an old nemesis. The ex-boyfriend of a former flame. A man who makes me angry and queasy all at once. Worse, he is a crack cattle jockey with a sharp eye—he makes a living buying and selling livestock, and is utterly at home in the sale barn. The only thing worse than meeting a man you despise is meeting him in his triumphal arena with your daughter at your side. I cannot tell a lie, I am suddenly happy no feeder pigs are available. I have every excuse to scuttle on out the door and back into the light. I thank Kenneth for his time, bid him good-bye, and walk back to the truck with Amy. On a farm not far from here, I have seen a sign: "Pigs for Sale."

I start the truck and we head that way.

The "Pigs for Sale" sign is still up, but there is no one home. I call the number on the sign. A man answers. No more feeder pigs, he says. Sold out. But try the guy over there on Randall Road. We drive on over. The farm is well kept and tidy. A man is mowing the lawn. "Guy up the road said you had some feeder pigs," I say after he shuts the mower down. "I do," he says. "I was just gonna send 'em to the sale barn tomorrow."

We walk into the barn through a passageway beside the milk house. There is a doghouse at the entrance, with a big old coonhound sitting at the door. He is secured with a heavy chain, but seems friendly enough, so Amy and I stop to pet him. He wags his tail and licks my hand. The barn is as neat inside as outside. The walk is limed, the farrowing stalls are clean, and the watering system is neatly plumbed. A good setup. Amy spots a sow with a litter of teeny piglets and naturally shoots right over there. "Oh, they're so *cuuute*!" she says. The feeders are in a pen on the other side of the barn, maybe six or eight of them, vigorous and alert. "Whaddya wantin' for 'em?" I ask, trying to sound all farmerish and hip. Inside, I am ridiculously nervous. Cripes, I've never bought livestock before. I wouldn't know a good pig from a bad pig if you hit the highlights with a laser pointer.

"I'm thinking forty-five bucks apiece," he says. Nervous as I am, I have been checking the market reports lately and know he's right in line. And if the state of the operation is any indication, these are fine pigs.

"I'll take two," I say.

I back the truck around to the passageway, which reminds me of the tunnel leading to a football stadium. The farmer has stepped into the pen and begun cornering pigs with a wooden door, holding it in front of him as he advances until he has one trapped in the triangle. Good in theory, but they are zippy little critters, and it takes some grabbing and lunging before we get the first one.

We each grab a hind leg, carrying the pig down the walk and out the passageway head-down. The moment a pig's hooves leave the ground it screams as if it is being scalded and will not stop until it has all four feet planted on a firm surface. In lieu of side racks, I have bent a cattle panel into a U-shape and secured it in the back of the truck with bungee cords. Hoisting the pig up to the tailgate, I am just reaching to lift the cattle panel when I feel a gigantic pinch on my butt, followed immediately by the sense that a great weight is hanging off my back pocket. At first I am so busy wrassling the pig, it doesn't register. But then the weight combines with the pain to buckle my knees, and I look over my shoulder. What I see is that hound—now transformed into a slavering Baskervillian meat grinder—masticating a Double Whopper's worth of my left butt cheek.

I utter an oath. One of the big ones.

Then I reach back and punch the dog in the nose. Hard. I have to use my left fist because I am fighting to hold the pig hock in my right. I smack the dog again. And then again, even harder. My fist is pistoning. Finally he turns me loose. I go right back to wrassling the pig. The second we get her inside the panel she goes quiet, snuffling at the bed liner like she's been there all afternoon. For all their screeching, pigs have a remarkable off switch.

My butt feels like it got sent to the laundry and run through a pressing mangle. It hurts so bad I can't walk right. The farmer is looking at me quizzically. "Dog bit me," I say.

"*Whaaat?*" In all the pig-scuffle, he didn't notice. "He's never done that before!" says the farmer, and based on his look of genuine dismay, I believe him. I figure it was us hauling that screaming pig past his nose that got the dog worked up. Probably triggered some primal killing neuron. Confronted with a stranger pilfering a protesting pig, the dog just snapped and went after the most prominent target.

Obviously embarrassed, the farmer helps me load the other pig. We skirt the dog widely. My butt has developed a bone-deep ache, and I hitch my giddyup to avoid contracting the glute. I nonetheless manage to keep up the small talk as we go around to the front of the truck and complete the transaction. Using the hood as a desk, I write out the check. I ask the farmer if I need to worm the pigs. He says he would. I ask him what kind of feed he's using. He tells me and fetches half a bag to get me started. I write the check out for an extra five bucks, and we're on our way.

The pigs ride home easy. The cattle panel works perfectly. I look back several times expecting they will be alarmed or skittering around, but they are riding happily, their snoots angled up and out to take in the view, their ears flapping in the wind.

When I get home, the butt pain is unmitigated. By craning my neck I can see tooth holes in the canvas, but no blood, so I go about unloading the pigs. I have been told they are remarkably adaptable animals, and they are proving it. Backing the truck up to the pen, I go after some wire and fencing pliers, and by the time I return they're snoozing like they've never had such soothing accommodations. "Can I mark them?" asks Amy. At first I don't see what she's getting at—then I recalled her helping Grandpa Bob mark the lambs. "Sure," I say, and she runs off for her carton of sidewalk chalk. She is back quickly, scruffing the chalk across their backs so now they each have pink and green stripes. I have to grab them by the back legs to lower them from the truck and they screech again, but go quiet as soon as they make contact with the turf. Scuttling off, they stand motionless in the shoulder-high burdock, grunting quizzically, first one and then the other, back and forth, as if they are having a conversation. Amy points to the one farthest away, a barrow. "That one's Wilbur!" Then she points to the gilt. "And that one's Cocklebur!"

Old-timers will tell you it's a bad idea to name your butcher animals. I lower myself gingerly down to one knee—my hinder still feels like I sat on a sea urchin—and make sure we have eye contact.

"You know why we have these pigs, right?"

"Yes?" There is a little question in her voice.

"In October we will butcher them. We'll cut them up like we do the deer. They'll be our food. It's OK if you name them, but remember they are not pets."

"That's OK."

I hope so.

The female lowers her nose first, scooping tentatively at the dirt with the ridge of her snout. When she raises her head, she is balancing a tablespoon dollop of soil above her nostrils. And this is the trigger. Both pigs drop their heads and begin scooping dirt wholesale. The innateness of it is fascinating; all their young lives spent on a grate or concrete, and given five minutes with the earth, they tuck into it as if born to it—which of course they are. Amy and I watch them with delight as they snuffle and burrow. At one point the female roots her snout deep into the earth and plows straight from one side of the pen to the other. She turns around. Surveys her work for a moment. And then, with an all-or-nothing flop, she drops lengthways in the furrow, rolling back to rest on the cool dirt, blinking with satisfaction at the open sky.

I wander up to my office in an attempt to get some work done, but I keep rising from the desk to gaze out the window at the pigs, like the kid at Christmas who keeps returning to the garage all afternoon to verify that the shiny red bike is *really, really there*.

It's hot out, and I'm worried they won't drink, so I walk back down and tweak the valve a few times so water drips in the dirt. I'm hoping they'll smell the moisture and get the idea. I'm also not sure how to introduce them to the feeder. I got the pig feeder free from my brother John. It's basically a tall, rectangular galvanized box with a roof-shaped lid. The lid tilts back so you can fill the box with feed, which then spills into troughs on either side courtesy of gravity. The troughs are covered by a series of segmented trapdoors. The pig merely noses the lid up and out of the way to eat, and when the pig leaves, the door drops shut to protect the feed from rain and small varmints. At first I prop the trapdoors open, but when the pigs nose in, a couple of the doors bang shut, causing the pigs to squeal and bolt. Eventually I open just one trapdoor, and when they get snooted in and start eating, I lower the lid gently on their brows. When they pull out, I open the door and repeat the process. After about three tries, one of the pigs raises the lid without assistance, and from that point on the buffet is open.

When I stop by the next time, they are snuffling inquisitively at the watering nipple. Finally one pig accidentally bumps the spring-loaded pin and a few drops of water release. Pouting her lower lip, she catches a drip. Then she noses the pin again. On about the third try, she opens her mouth and clamps it over the nipple, releasing the water to flow freely down her gullet. Soon they are taking turns at the nozzle.

I return to the office. I manage to get a little work done, but I have to lean forward to keep the pressure off my throbbing hinder. By suppertime not only has the throbbing failed to recede, it has developed a specific rhythm, at which point it strikes me that if your average cogent person found a loony bluetick coonhound dangling off his fanny by its four main teeth, he might have already taken time to inspect the damage.

I toddle off to the house.

Alone in the bathroom, I back up to the mirror and drop my shorts. And what I say aloud is, *"Holy Shnikies!"*

The greater portion of my left butt cheek is obscured by a hematoma the size of a personal pizza. The hue of the relevant skin is something along the lines of stomped blueberries. In a nod to symmetry, a quadruplicate set of puncture wounds brackets the bruise as neatly as the four cardinal points of the compass. First thing I think is, *I gotta SHOW this to somebody!*

I call Anneliese from the kitchen. She is used to indulging my fascinations—that is to say, the woman can stifle a yawn—but it cheers me to report that when she sees the bruise, her eyebrows shoot right up. I have her bring me the digital camera so I can get pictures. This I accomplish by twisting around and shooting at the moon in the mirror. You festoon up a blemish of this caliber, you want some documentation for the possible grandkids.

I dither over whether to call the farmer back to see if the dog's shots are up-to-date. I don't want to bother him or get him worked up. It seemed pretty clear to me that the dog had been whipped up by all the excitement. People who excuse biting dogs rank high in my pet peeve list, but I truly believe this incident was an aberration. Then again, there are specific drawbacks to rabies—or *hydrophoby*, as the cowboys in Louis L'Amour books called it. On the other hand, my puncture wounds are relatively superficial. In the end, I just tell Anneliese to pack me off to Urgent Care if I start walking into walls or slobbering. For legal purposes I should say that if you find yourself in similar circumstance, I cannot recommend that you follow my health-care decision-making tree. At dusk I check the pigs one last time. They are lying tight against each other, settled in for the night. I go back in the house and climb the stairs, kiss Amy good night, pause at the crib and listen to Jane breathe, and then crawl in beside my beloved Anneliese. As I pull the covers up and roll gingerly over to sleep on my unbitten side, I think, Yessir—we're in the pig business.

Jane has enough muscle tone now that I can prop her up in the old green chair across from my desk and write while she grins at me. It's pretty handy really—she can't crawl, so she's pretty much stuck wherever I stick her. She grins and slumps, and every now and then I give her a boost. I can usually get ten or twenty minutes in before her face clouds. Then I have a series of stair-stepped actions I implement to postpone the ultimate inevitable monsoon. First I turn the stereo on, normal volume. This captivates her for another five to ten minutes. Then when her brow begins to furrow, I crank the volume. She seems particularly soothed by Steve Earle's thumping version of "Six Days on the Road," which buys me three minutes and eight seconds of additional productivity. Finally, as a last resort, I pull out the guitar and sing some of my own stuff. I sit on the footstool so we are face-to-face. Jane grins and coos for about two minutes, at which point she finds my oeuvre derivative, and her lip begins to square off.

There is this stage—right when she is transitioning from *ver-klempt* to full-out bawling—when her lower lip widens and rolls out, but instead of a rosy little pout, it locks into a position so straight-edged you could grab the kid by the feet and use the lip to strike off wet concrete. One day when she was going from happy to sad I shot the entire transition on a digital camera just so I could document the geometric rigidity of it. I realized halfway through that I was behaving on a par with the heartless producer who films the crying kid cast to demonstrate inferior products in a diaper commercial, but I kept snapping anyway.

The day after the pigs arrive, Amy harvests two of the radishes that survived the excavations of Fritz the Dog. She holds them up to either side of her ears, and I take a picture. She is beaming, her front teeth still missing. Then she runs off to rinse and eat them. I remember doing the same thing at her age, rinsing a spring radish under the brass standpipe beside the garden. I remember the cold water made my knuckles ache, I remember watching the dirt dissolve and flush from the root hairs to leave them feathery and white; I remember the red skin shining beneath the film of water. I always nibbled the bland taproot first. And then that first full bite through the scarlet skin, the crisp crunch, the excitement of springtime snack food fresh from the ground. Over by our standpipe, Amy's lack of front teeth put her at a disadvantage, but she's gnawing assiduously, the radish jammed way back in a corner of her mouth so she can get at it with her molars. Her lips are pulled off-center, but I'd say the crinkle in her nose substitutes for a smile.

Now that the pigs are in place, I am going to get serious about building that chicken coop. As is standard procedure when I say "I" am going to build or fix something of any size or substance, there is a technical adviser/handholder involved—in this case, my pal Mills. Mills is a good man, but he has regularly led me down the path to iniquity—it was he who got me started on carp shooting, and I have lost a ton of man-hours in the endeavor. He also got me hooked on auctions, and we do spend a little too much time on a certain popular online auction site. My weakness is anything to do with my hometown of New Auburn, Wisconsin, or pretty much anything sporting a vintage International Harvester logo. As for Mills, he is constantly on the prowl for anything to do with firefighting. He has an astounding collection of antique fire extinguishers, and his driveway is lined with discarded hydrants.

Mills is especially valuable in an endeavor like building the chicken coop, because he has a lot of very cool tools—chop saws, nail guns, and so on—and he is quite handy. Even more important, he is a professional-class scavenger. I don't mean a guy who picks something up at a thrift sale now and then (he does), I'm talking about a guy who goes to nearly every auction within a forty-mile radius; is an eBay power seller; knows the guy in the basement of the local grocery store who has all the free five-gallon buckets; can put a word in for you with the guy who handles all the scrap wood from the furniture factory; and—this is huge—is on a first-name basis with the *dump guy*! Mills owns a farmstead. His red barn is jammed with every conceivable form of potentially useful scrap and geegaw—steel barrels, discarded RV siding, plumbing supplies, secondhand plywood and discontinued signage, doorknobs, hinges, and used Styrofoam sheeting. Some of the best stuff is outside, hidden from sight behind the pine trees that ring the property. Mills calls these stashes his "Sanford and Son piles." Treated posts, barrels, trailer frames, angle iron—you name it, somewhere out there in the brambles beneath a tarp, he's got it.

The other day I introduced Mills to Craigslist, and our relationship may not survive. Problem is, our geographic search parameters overlap, plus we regularly covet the same items. Having been on the lookout for a radial arm saw, I was excited when I spotted one on Craigslist for a most reasonable price. It was located south of me in the tiny town of Humbird. In the photo the saw was posed in front of a red garage and looked promising. I contacted the seller immediately. Too late, he said. Someone had already claimed it. Two days later, I ran into Mills. "This Craigslist thing—*woo-HOO!*" he said. (In conversation, Mills runs heavy to italics.) "I got an *amazing* deal on a *gorgeous* saw!" "Poacher!" I said.

Lately I have been scoring stuff from Craigslist nearly every week. Rain barrels, fence posts, lumber. I even managed to find another radial arm saw. It was a newer model than the one Mills stole, and I paid less. "*JINKIES!*" he said when I told him. Nowadays we regularly consult with each other before making contact on Craigslist items. It is my understanding that the original purpose of Craigslist was to help people in San Francisco locate apartments. I am tickled to think it wound up causing two knuckleheads in Wisconsin to fight over used barbed wire and secondhand pickle buckets.

Since Mills has all the equipment and most of the supplies, we decide it will be easier to build the coop at his place, prefab style, then haul it over to our place in pieces. So I am on my way to his house now, with Amy in her booster seat behind me. With my schedule over the past year, "our" efforts to homeschool Amy have quickly devolved into Anneliese doing all the day-to-day hard work while I provide the occasional off-kilter field trip—in this case a morning spent constructing a chicken coop in the company of two grown men whose greatest aspirations tend to center around finding any excuse to shoot arrows at overgrown goldfish.

"Where does Mr. Miller live?" asks Amy as we drive. "Mondovi," I say. "Is that a city or a state?" asks Amy. This is a recurring lesson. We keep two large maps on the wall of Amy's bedroom so she can track me when I call in from the road. Despite our efforts, Amy struggles with the difference between city, state, and country.

"Mondovi is a city," I say. "It's in the state of . . ."

"Wisconsin?" says Amy. Tentative, but correct.

"Yes. And Wisconsin is in what country?"

Silence.

"The United . . ."

". . . States of America!"

"Where does your daddy work?"

"Denver."

"Is Denver a country or a state?"

"A state?"

"No . . ."

"A country!"

The exasperation hits me immediately. I am ashamed at how hard it is for me to maintain my patience.

"No! Denver is a *city*. And Denver is in *Colorado*. That means Colorado is a . . ."

Silence. Then, quietly, " . . . a country?"

Tasting the dust of my molars, I make a mental note to shake Anneliese's hand when we return and perhaps print her up some sort of framable certificate honoring her persistence. She would actually prefer that I handle the spelling lessons for a day. The thing is, fifteen minutes later we meet Mills for breakfast, and as Amy jaws with him on a wide range of topics (central theme: horses), I feel the exasperation dissipate in the face of pride in how she comports herself—politely, and with poise beyond her years. It is good for me as a parent to see her with a little rope to run, a chance to operate in the big world with the skills she has. Cities and states can wait. Worse comes to worst, she can carry an index card for the rest of her life.

Amy calls Mills "Mr. Miller" because that is what I call him in her presence. I am old-school in this regard. I believe it benefits the child to know who the grown-ups are. "Heavy on the 'Mister,' " my dad used to say.

When we arrive at Mr. Miller's house, it looks like the place has been preset for a shoot with *This Old House*. Several work-tables are arranged on the blacktop drive in front of the red barn, and each is stocked with a wide range of saws, hammers, screwdrivers, drills, nails, screws, earmuff hearing protectors, safety glasses, and a generous selection of fitted work gloves. The chop saw is on its stand and plugged in, the air compressor that powers the nail gun is all set with hose neatly coiled, and in a truly elegant touch, bottled water is chilling in a cooler.

First thing I do is strap on my tool belt. Gosh, I like tool belts. Just the very look of them confers competence. I like the way the belt hangs gunslinger low and loose, the hammer dangling in its loop, the handle gently tapping at my thigh as I walk. I like the heft of the nail pouch at my hip, and the way the big fat tape measure slips neatly into its special pocket. I tend to overdo it on tape measures. At last count, I owned eight of them. But the thing is, you're forever needing a tape measure for this little project or that, and my level of disorganization is such that the only useful countermeasure is to throw one in the cart on every other trip to the hardware store and just sow them willy-nilly all over the place. At this very moment I have two in my office, one in my car, a pair in the house, and at least three in the shop.

I fit Amy with padded kneelers, safety goggles, and work gloves, and then hand her a hammer. She grins when she hefts it and looks around for something to hit—evidence that while variations persist, the love of gear crosses genders. In truth, part of the lesson we hope to convey today is that girls can build chicken coops, too. In Amy's case, the lesson will be redundant: when the light fixture in my bathroom needed replacing, my mother-in-law—she who supported her children by climbing telephone poles for twenty years—did the job because bare wires leave me frightened and confused. She also put the phone line into my office. Amy's grandmother in Colorado raised five kids and ram-rodded the farm for twenty-seven years after her husband was killed in an accident. By way of contrast Amy has watched me struggle for twenty minutes to get two corners of a four-sided cold frame to match up. The explication of gender roles is all well and good, but it is likely my hand in this will be light. (Although as a fellow who put himself through nursing school by working as a cowboy in Wyoming, I have addressed the subject previously.) I do anticipate a time when I will have to explain to Amy that while most men are happy to see a woman in a tool belt, it is sadly for all the wrong reasons.

If the coop project is to go well, it will all come down to Mills. I am a loyal laborer, and will pitch in full bore, but even with proper guidance I tend to run off the rails. Part of it is a patience issue. Once I get started, I want to finish. This leads to rushing and improper material usage, to say nothing of improper application of hardware—say, trying to drive finish nails with a plumb bob. And even when I do slow down and read the directions, things have a way of going wrong. Remember that electric fence I hooked up for the pigpen? I did the whole thing exactly right—spaced and sank three grounding rods instead of settling for just one, linked them together, and clamped (rather than just wrapping) the wire as indicated . . . a month passed before I went to open the shed door and discovered that I had run the ground wire in such a way that the door couldn't slide on its rails without cutting the wire in two. If life was a state fair, I'd have a giant shoe box full of green ribbons embossed with the word PARTICIPANT.

I want the chicken coop mounted on skids, as we intend to

move our chickens around. Also, because it does not sit on a foundation, it is not viewed as a permanent structure and will therefore not be taxed as such, or so I believe until the assessor tells me otherwise and I pony up. Since the skids will be in direct contact with the ground, I tell Mills they need to be made of treated lumber. He grins. "Come with me!" Dressed in his sleeveless T-shirt, ball cap, white athletic socks, and Crocs, he leads us up a trail into the pine trees off the side of the yard, past several Sanford and Son piles, and then, with the civilized flourish of a sommelier pulling back the velvet curtain shrouding a particularly pricey corner of the wine cellar, he strips back a tarp to reveal a stack of beautiful green-treated six-by-six timbers that will be perfect for the job. We lug them back to the yard, saw the ends off at an angle, and begin framing up the floor.

I try to involve Amy wherever I can—when we trim the ends of the skids, I show her how to use a carpenter square to draw a pencil line at the proper angle, and in between, how to stow the pencil behind her ear. Because the skids have to be the same length and we have four six-by-sixes to choose from, I give her the tape measure and let her find the two longest, then determine how much we have to cut from the longer of those two to make them the same length. This leads to a discussion of inches and feet and how when you write measurements on a scrap of board, inches are denoted with a double hatch and feet with a single. When Mr. Miller fires up the saw, we put on our earmuffs and afterward discuss the importance of hearing protection. When we make a mistake, I show her how to pull a nail, and I show her how to extend the reach of the hammer claw by putting a shim beneath the head. Once we get going on the deck, it is mostly a matter of driving straight nails into flat boards, so she can really go to town. She whales away at a steady pace, bending a nail now and then but just as quickly pulling it and grabbing another from the plastic Folger's can. To make it easier to hit the underlying frame I show her how to use a chalk line, and of course she loves this—snapping the taut string with a cottony twang and watching the elongated cloud of purple chalk dust float away and dissipate on the breeze, then reeling the line in to recoat it with chalk, just like a fishing reel with no pole.

We work into the afternoon. I try to keep teaching without being overbearing. I let her measure and mark the boards to be cut. I give her little problems, like, if we need one board sixteen inches long and another board two feet long, can we cut them both from this one long board? I find myself experiencing none of the frustration I felt during the city/state/country grump-up, and Amy takes the lessons well. But mostly she holds her hammer in both hands and haves at 'er. Perhaps the finest thing I teach her all day is how to keep a couple of extra nails in easy reach by holding them in your lips. She loves this, and is currently well-suited: the nails fit nicely where her incisors should be.

At quitting time we have finished the deck. It doesn't look like much—just a wooden floor on two large skids. But it's a start. Mr. Miller took our picture just before we finished. There's me, a lumpy bald guy in cheap sunglasses with sweat darkening his T-shirt collar, resting my hand on the shoulder of a gangly little gap-toothed girl in shorts and pink Crocs, her head higher against my sternum than even a month ago, squinting in the sun and quite literally standing on a good day's work, and—I hope—on a little piece of her education.

Back home, I am walking down to check the pigs when a press of cold wind rushes the yard behind me, and when I turn to look back my heart startles, because a towering billow of pollen has spun from the pine crowns and is twisting up and over the house, so thick and yellow I actually think for second that the attic is afire. Majestic and surreal, right out of the blue. And then it is gone. Down in the pen, the pigs squeal and zigzag madly, kicking up their heels as the first drops hit. Now the wind is on a straight line, and the space between the house and granary goes white as it scours a blizzard of dandelion fluff from our overgrown yard. Then the real rain hits with its hiss and splatter, driving the pigs to their shelter and the dandelion fluff to ground. The land is dry, dry. Our yard is like a brick. We need this.

It rains hard, but not long. In the wake, the sun is already poking through, and steam rises from the asphalt by the garage. A rainbow forms across the ridge. Amy is spinning across the yard with an umbrella. Just like she did, she tells me, "when I was a kid, and I was three."

You can really go off the rails with this scavenging business. While working on the pigpen as the earth has warmed, I noticed a number of seedlings cracking the dirt. Their cotyledons were fat and spoon-shaped on the order of a squash or melon. I assumed the previous owner must have tossed some garbage down here and figured what the heck, they're off to an early start, I'll transplant them. Take them as a gift of the good earth. So over a period of a week I spoon out the sprouts as I find them and place them in a careful row along the far side of the garden. Soon the first real leaves emerge. They are pointy, kinda like you might see on zucchini. I begin to get a little nervous, however, when I start seeing the things popping up all over the barnyard and around the outbuildings. Then while clearing out the pigpen tangle, I notice a pattern in the distribution of the sprouts and put two and two together: I have been transplanting wild cucumber. This is the equivalent of transplanting thistles. Honestly, I should get a plant book or something.

Anneliese is taking the lead on the garden. I helped plant onion sets and some kale, but she is doing most of the rest of it. So far she has put in turnips, chard, more radishes, two rows of tomatoes, and several hills of potatoes. And in a touch missing from my bachelor gardening days, she plants marigolds at the end of each row.

Lately Jane fights her bedtime with a ferocity that easily outsizes her frame, and we have fallen into a pattern after supper in which Anneliese gardens in the remaining light while I try to settle the baby. I am often surprised to find myself here, holding this teensy howling beast to my chest, catching the scent of baby powder, and contemplating how I have come to understand what a "onesie" is at this late stage in my life. Here I am buying diapers when most of my contemporaries are buying graduation cakes.

The kid can really holler. People say that, but seriously: when I cradle her to my chest, invariably she'll hit a note so pure it triggers my tinnitus—the ear nearest her mouth damps down and rings long after she is snoozing. For a while I did my best to ease her gently to sleep in a rocking chair just the way they do it in fabric softener ads, but the screeching went unabated. Then one desperate evening I sat on a giant rubber yoga ball Anneliese keeps in the bedroom and started bouncing. The baby's cries softened. I bounced higher. The cries got softer. One does not wish to do harm, so I held her tight and close, steadying her neck and head in my palm, and went full-bore pogo-butt. I'm talking lift and clearance. Nothing gentle about it. And in five minutes, she was asleep. Now we bounce every night. Our bedroom window overlooks the garden, and for the rest of my life when I think of our first year on the farm I will remember my baby clamped to my chest and my beautiful beloved wife grubbing in the garden at twilight, working diligently to feed us over the seasons to come, my vision of her springing in and out of frame with every bounce.

Soporific bouncing is not just for babies. I am looking at a photograph Anneliese took three nights ago. My feet are on the floor and my butt is still on the ball, but I have tipped over backwards on the bed. I am sound asleep and so is Jane, curled like a little possum on my chest, my hand still across her back as it was when I drifted off feeling her breath rise and fall.

The pigs have eaten the bag of feed I got from the farmer, so today Amy and I make a run to the feed mill in Fall Creek. There is much that is similar to the New Auburn feed mill Dad patronized when I was a kid—the loading dock, the attached office, the dusty hand truck, and feed pallets all about—but the operation is much bigger than the one of my childhood, with towering bins and a radiating tangle of augers. Dad used to shovel our corn and oats into a howling subterranean grinder at the front of the mill, and then a few minutes later a man named Big Ed brought it back out the door in heavy bags that we wrassled into the truck bed. Today when the man wheels out our pig feed it is pre-bagged in paper sacks laced shut with pull strips, but when I stand at the edge of the dock and sling them in the truck bed, the soft heavy shape of the feed in my arms triggers a comfortable muscle memory. Inside the office the man rings us up on a computer rather than a notepad, but I am pleased to see a pair of farmers lingering and telling lies, just like in the mill of my childhood.

The New Auburn feed mill is long gone. It changed hands several times, the farmers disappeared, and I was a member of the fire department when we burned it to the ground for practice. As we pull away from the Fall Creek mill, I tell Amy how after the New Auburn mill closed Dad used to go to the Chetek mill, where instead of shoveling the feed from the pickup he backed up until his front tires clunked into a bracket and then the man inside hit a switch and a winch lifted the whole front of the pickup into the air, tipping it higher and higher until all the corn and oats just slid right out. I can't imagine such a thing is allowed now, but back then we kids were allowed to ride in the cab as it rose in the air. Amy's eyes are wide. "Oh! Can we go to *that* feed mill?"

On the way home we stop at the post office to mail a package to a friend serving a tour in Iraq. Amy drew him a picture and a note about the pigs. She asks, and so I try to explain war. As we drive through the tunnel of trees shrouding the last hill before our house she says we sure are lucky to live here. She doesn't know it, but that boy in Iraq just lost some friends one Humvee ahead of him. I say, Yes, baby, we sure are.

A week now with the pigs, and so far so good. I'm still eager to get the chickens, but for now the pigs are a terrific diversion. It's fun to take the slop buckets down and watch as they devour every single table scrap and leftover and rind and trimming and old potato we cannot use. The rubber pan I bought is proving to be pretty much useless, as they wade right into their food and upend it in short order. At least it's durable. They root and worry everything in sight. It took them just two days to work loose the legs supporting the water barrel to the point where it was teetering. I had to string electric wire around it to create a boundary and keep them from knocking the whole works flat.

I'm in over my head, but if I pay attention, they give me hints. On one particularly hot day I eased down to watch them and found both pigs at the spigot. They were lain draped across the ground, raising their snoots just far enough to nudge the steel nipple and release the water. Each in turn would take a mouthful, then let it dribble slowly to the ground. Pretty soon they had moistened a good patch of dirt. They rooted around at it, stirring it with their noses. Then they dribbled more water and stirred it again. The cycle continued for quite a while until they had dug a muddy bowl-shaped hole. Soon the hole was so deep they were able to get beneath the new electric strand and were again threatening to undermine the posts supporting the barrel. I made a note to reposition the spigot over the concrete bunk next year so they can't excavate.

Not a bad idea, but not the main point. When I told Anneliese what they were doing, she clarified the obvious. "They need a wallow," she said. Of course. Pigs can't perspire and they need a wading pool to keep cool. I went back and hosed down one corner of the pen. I am careful not to spray the pigs themselves. I have read that the shock of cold water can give them a heart attack. Soon, however, they are scampering in and out of the hose stream, reveling in the cool and snouting around in the drenched dirt. They show no ill effects, and before long I throw caution to the wind and train the water directly on the pigs. Wilbur grunts and just stands there, but Cocklebur actively seeks the stream and often blocks it from Wilbur as she lets it play over her nose and into her mouth. When I finally close the hose their undocked tails spin a happy whirligig as they nuzzle and roll in the fresh mud.

During one of my fits of activity, I built a shelter to provide them protection from the sun and rain. I began with a fine vision of what the shelter would look like. I even planned to roof it with some used shingles I found down in the shed. Nothing says redneck like a blue tarp roof, and I swore I wouldn't go that way. As usual, I overdreamed and underbudgeted, and wound up banging together a bunch of castoff two-by-fours, several chunks of warped particleboard, and—due to hit the road for a stretch with no time for shingling—finished it off with a nice blue tarp. Sigh. On the bright side, it will be easily spotted by the assessor and should depress our property values accordingly.

The pigs have so far disdained the shelter, and as a result their ears are badly sunburned. Not my fault, I think, but perhaps a better farmer would slather them in SPF 40.

A while back our neighbor Ed drove up the hill with his tractor and rear-mount tiller and churned up a patch beside the pigpen. I planted several rows of sweet corn, some zucchini, and broadcast a pailful of soybeans Amy and I shucked on the porch steps. The plan is to feed the pigs zucchini and sweet corn and then eventually turn them loose on the soybeans and everything remaining. In the process we're hoping they'll chew up the ground and give us a nice garden plot for next year. Piggy as rototiller.

Ed came up because when I tried to till the sod our little tiller hopped and bounced and barely scuffed the dirt. Ed's machine did the job in a trice, and he wouldn't take a thing in payment. I am grateful for the help, but even more grateful for the spirit in which it was offered. It sounded like he was hitting some rocks down there and I cringed to think what he might be doing to his equipment.

Taking a break from the desk one afternoon, I put Jane in the backpack and take Amy down to check the pig patch. Everything has come on nicely, but because I scatter-sowed the soybeans, I can't weed them properly, and they are succumbing to quack. We are wading through the mess when five feet ahead of me I spy a female pheasant and a scattering of pheasant chicks. They are huddled at the edge of an open spot where the previous owner of the farm had a burn pile. The entire family is utterly frozen and pressed flat into the sand and ash. Perhaps the black ash is warm in the sun. What caught my eye was the mother's own eye blinking. When I lean in just the slightest bit for a closer look, the mother flinches, ducks her head, and nearly bolts, but in the end she holds. *"Look!"* I stage-whisper at Amy, then, *"Don't move, don't move!*" A look of alarm crosses Amy's face immediately, and I whisper, "It's OK, it's not a skunk or a bear, look, baby pheasants!" Even from five feet it takes Amy several hard looks to spot them, but when she does, her face lights up. We study them silently. How fragile this all is, the mother with her fuzz-balls and coyotes, fox, mink, and fishers all about. "I want to hold one," Amy whispers. I explain why we must leave the birds be, and she is satisfied to leave.

We are still mincing softly away when the pigs break into a fit of oinking and goofball galumphing. I recall how Mister Big Shot was haranguing us the day we worked on the pen, and I wonder if perhaps he was being territorial because this brood was about. Perhaps the old boy was more than strut and cackle . . .

I get back over to help Mills work on the coop again. There are the usual mishaps. I painstakingly craft two tiny chicken doors. They are hinged on the bottom and designed to drop open, form-ing miniature ramps. I even cut and nail a series of little cleats the full length of each ramp so the chickens won't slip and fall. Problem is, I get things backward in my head and build them too wide. An oxymoronic bout of fine-tuning ensues. Mills gig-gles, which helps take the pressure off, and I don't throw a single tool. In the meantime, Mills is constructing walls. He's working steady, the automatic nailer firing with a hiss and thwack as the nails are driven home. I am due for another long stretch of road time and won't be back for a while. I know Mills enjoys projects like this and will probably continue in my absence. Somewhere in my subconscious or shallower, I'm banking on it, in fact. I should be a better person.

The baby continues her bedtime protestations and has been right up to the edge of colicky. One night when Anneliese is in the garden and I am bouncing on the ball and nothing is working, I try humming the standard Brahms lullaby. The kid rages on unabated. Drowns me out. So on a whim, I begin singing the lullaby really, really loud. *"LA-LA-LAAAAAH, LA-LA-LAAAAH, go to SLEEEEEEEP NOW MY BABY!"* and by jiminy it works. Shocked her into stopping, I suppose. I feel like Papa Axl Rose.

You can't holler lullabies in the deep of night, however. When she wakes crying I bounce her on the ball in the dark, or walk the floor, but mostly it comes down to Anneliese nursing and rocking her. Lately when I sense that some well-meaning mother is about to give Anneliese advice on how to get the baby to sleep, I jump on the conversation like I'm smothering a grenade. Whatever it is, we've tried it, and it hasn't worked. And the teething hasn't even begun.

Here I am set to leave again, my wife so tired, and so much undone. Again I look at the unmown lawn, and for the thirty-seventh time I tell Anneliese I plan to fence the yard and get some sheep. Let them eat the lawn and sell them in the fall. Save on gas and mowing time. Anneliese has not uttered a word of complaint about my absences, but now she looks at me.

"About the sheep," she says.

"Yes?"

"No sheep."

Later that evening she shares her line of thinking. "I have this vision of you in Des Moines, talking about writing and raising sheep—meanwhile, I'm running through the brush with a howling six-month-old under one arm and dragging a bawling seven-year-old behind me with the other arm while we try to get the sheep back inside a hole in the cobbled-up fence."

This is very hard on my pride, and pretty much on the money.

Sheep. Maybe next year.

And yet there are beautiful days. On a lovely Saturday morning when my mother-in-law and all three sisters-in-law are visiting and request some grown-up girl time, I put our two recumbent tricycles in the back of the pickup truck and drive to a local bike trail (if you question the environmental propriety of trucking one's bicycles around, I encourage you to attempt a series of 10 percent grade hills on a tandem recumbent with a weeping seven-year-old as stoker and get back to me). Amy is pouty upon departure, wishing as any little girl does to be one of the big girls, but by the time we hit the bottom of the first hill she is happily shooting the breeze. This is becoming an established pattern. After a short drive to the trailhead, we unload the trikes, hook them in tandem, and set out. Behind me she narrates nonstop. "Can you tell I'm helping?" she sings out when we hit a slight grade, and indeed I can. The poor kid, as tall as she is, is nowhere near sized for this bike and is basically lying flat so she can reach the pedals. We roll along the river into downtown Eau Claire, then cut across the old railroad bridge and down to Phoenix Park, where today a local arts festival is in full swing. As we draw near we can hear live music. Amy stops pedaling and sits up in her seat. "It's the *Cheese Puff Song*!"

The "Cheese Puff Song" has been in heavy rotation around

our house for some time now. The artist, Magic Mama, is a local resident. I hustle to get the bike parked, and Amy makes it to the music tent in time for a chorus. For the rest of the show she sits glowing in the front row, singing along to the songs she knows—including "Go Barefoot" and "Take It Outside"—and happily participating when Magic Mama hands out used potato chip bags and encourages the children to crackle the bags in time.

Next we go to the craft tent and make a puppet. While Amy swabs glue and cuts out eyeballs, I take two thin strips of construction paper and show her how to make dangly accordion arms like I learned back when I was eating paste. When the puppet is finished, we wander through the farmer's market and stop at the local foods booth, where our friend Aaron lets us sample farm-direct apriums. Amy spots a bevy of belly dancers and says she wants to watch. Who am I to argue? Amy likes the belly dancers very much, and points out her favorite costumes. In order to serve honesty I must skirt the edge of propriety and report the demonstration expanded my appreciation of the female form in both an artistic and a more basic sense, and it didn't hurt that the scent of patchouli was prevalent throughout. Even as a guy with pickup truck sensibilities, I have always gone a little weak in the liver for patchouli. After the belly dancing we find the body art tent. Amy gets a henna tattoo on her foot and I get a henna wedding ring. Can't lose it. Later we wander back over to the music tent, where the musician Bruce O'Brien is accompanying himself on banjo while singing another one of Amy's favorite songs, the chorus to which goes "Peace and joy and harmony, and love is in the middle." She is sitting beside me on a hay bale, and when she leans her head to my shoulder during the chorus I hope he'll reprise it again and again. After the concert Amy asks if she can talk to Mr. O'Brien, and when she looks up at him without guile and says, "I really liked your music," I get teary at her earnestness. We circulate a while longer, Amy gets a ride on a goofy bicycle that is doubling as art, she plays a while with the children of some friends, but there are clouds moving in now, so we have to go. On the ride back Amy pedals just as eagerly as she did on the way in, and when we are tooling right along she says, "I thought this day wasn't going to be fun, but it *was!*"

I have been saving the best surprise for last. Anneliese and I have split an order of chicks with our friends Billy and Margie. The chicks have arrived, and I am taking Amy to meet them. (We have a week-long family trip planned soon and won't take our chicks until after we return.) When we arrive, Billy and Margie lead us into the garage, where the chicks are being kept in a wading pool lined with wood chips. They are a sprightly, multicolored bunch, warm under the heat-lamp light. Amy peers over the edge at them, and immediately her eyebrows knit, not in a frown but in that universal feminine look of care. "Oohhh," she says. "Can I hold one?" There was a time Billy doubled as a bartender and bouncer in the type of taverns you enter through a small dark door. He rode a thunderous motorcycle and had the size to back down any man. Today he remains an imposing figure, but his spirit is gentle and he engraves tombstones for a living. A man who has gone from bare-knuckle to Bashō, he seems the perfect fellow to carve the dates of your birth and death, so wide is the breadth of his understanding. Now he reaches into the blue swimming pool, carefully closing a chick within the cavern of his roughened hands, and passes it gently to Amy, his blackened nails and shredded calluses (he is recovering from the peak of the carving season, when he catches up with a winter's worth of graves in time for Memorial Day) of another creature in contrast to Amy's soft white palms and slender fingers open wide to receive the bird. Carefully she closes her hands until the chick is cupped within, then draws it to her face and inclines her cheek to its fuzzy head.

CHAPTER 7

My daughter is weeping in the timothy. She is a sad sight with a sparse handful of stems dangling from one hand, grass clippers dangling from the other, head tipped back as she beseeches the sky. From my perspective—framed by the window over the kitchen sink—what we have here is a scene composed by Andrew Wyeth and retouched by Edvard Munch.

The girl is weeping in part because I am a cheapskate.

Among the trove of supplies and accessories provided by Aunt Barbara when we took possession of the guinea pig back in January was a neatly sealed plastic bag of prime timothy hay. Every day when Amy replenished his tiny hay rack, the creature tore into it eagerly, sometimes whistling with delight at the first sound of crinkling plastic. When the original bag was nearly depleted, I stopped by a local pet food store for another. Wanting to maintain the standards of quality established by Aunt Barbara, I searched the racks until I found the exact same brand of timothy and grabbed a 12-ounce packet. I've cut and stacked a lot of timothy in my day, and while carrying the bag to the checkout I was admiring the quality of the product—a weedless sheaf of fat-leaved stalks all dried to a uniform pale green. Really top-shelf stuff. Then the woman at the register swiped it across the bar code reader. When the price popped up, I suddenly understood what was making that guinea pig whistle. I made a very similar noise, although it quickly tapered off to a wheeze.

Numerals are not my thing, but sometimes one must quantify astonishment: Beginning with a generous interpretation of current Midwestern market prices as provided by the county extension agent's Web site, the finest prime grade hay will run you somewhere in the neighborhood of $175 bucks a ton. That twelve-ounce packet of guinea pig hay rang up at $6.98. Rounding *down*, that's 58 cents an ounce. Tappety-tap, there are 32,000 ounces in a ton. Times point-five-eight, equals: the stuff I was carrying across the parking lot to the van costs $18,560 per ton. Next time I rent an armored hay wagon, I remember thinking as I scanned the lot for grass bandits. I briefly considered selling the guinea pig and all his toys, renting a safe deposit box with the proceeds, and stuffing it with hay that I would then roll over into an individual retirement account. Instead I crawled into the van, locked the doors, took my cell phone in trembling hands, and called my father. Having recently heard him apologetically report that he was selling organic horse hay for upward of $120 per ton, I wanted to tell him these horse people are pikers, and guinea pig hay is where it's at. Sell everything, I told him when he answered, and get yourself a miniature baler. Before I drove home I secreted the timothy beneath the spare tire and conspicuously placed my open wallet on the dashboard so that should I encounter highwaymen they would go for the wallet first.

"Tell Mr. Guinea to enjoy this," I said as I handed the bag to Amy. "That'll be the last bag." She looked at me quizzically. Because the guinea pig is serving as training wheels for a future alleged horse, I thought it might be helpful to explain my reasoning through parallels drawn from the equine world. "The finest horse hay in the land costs 175 bucks a ton." I was in full royal declarative mode. "The stuff we're feeding that guinea pig costs $18,560 a ton!"

Seven years old, and she hesitated perhaps two nanoseconds to read the seams on the ball before smacking it straight back at the pitcher.

"So we should get a horse."

Occasionally one is provided glimpses of the road ahead. I am hoping there are rest stops.

Still, for now I am in charge, so when I noticed patches of volunteer timothy sprouting in our overgrown lawn and out on the ridge, I did rejoice and sent forth my daughter to gather stalks together. Before handing off the clippers, I placed a hand on her shoulder and patiently explained the dynamics driving this decision. Before the monologue concluded, I had invoked principles of self-sufficiency, economies of scale, the comparative nutritive value of native grasses, footnotes from a nice little chart available through the county extension office, and—for zip—the fable of the grasshopper and the ant.

Amy found this unconvincing. So then I tried explaining it *im*patiently, and now there are tears on the lawn. Certainly I am economically justified in sending my seven-year-old out to harvest grass; one can additionally argue the case along the lines of physical exercise and personal responsibility and further defend it as a proactive move to ensure she gets her vitamin D. It is also possible that the poor girl is suffering the projections of my own fond memories.

I loved making hay.

Of course you don't *make* hay, and in fact the only time we ever used the phrase was in the metaphorical sense: *Gotta make hay while the sun shines!* When my dad picked up the phone to call his friend and neighbor Jerry, he'd always say, "You gonna *bale* today?" And if the answer was yes, you also knew "You" meant "We," and you went to looking for your haying gloves.

In the early days Dad cut hay with a simple sickle bar mower—basically a seven-foot rolled steel plank fitted with rapidly recip-rocating blades. You can get the idea by placing the palm of one hand over the back of the other, fanning your fingers, and shuf-fling the top hand back and forth.

Dad ran the mower off the back of his small Ford Ferguson, where it could be raised and lowered by a set of arms extend-ing from the tractor. The power was supplied by a splined shaft (called a power takeoff, or PTO) that protruded from the back of the tractor and spun a flywheel attached to a pitman bar. I was always captivated by the pitman linkage because it reminded me of the linkage I had seen on steam locomotives in the cowboy shows we watched at Grandma's house. One end of the pitman bar was attached to the outer edge of the flywheel and therefore followed the circular path described by the flywheel. This caused the other end of the bar, which was flexibly attached to the sickle, to plunge back and forth, making the sickle bar do the same. When the tractor was operating at full throttle, the flywheel end of the pitman bar whirled to a transparent blur while the sickle end pistoned so furiously you expected it would yank the sickle in two. There was something magical about the way it converted rotary energy into linear energy—or, from a child's point of view, a circle into a straight line.

Before he set out to cut hay, Dad would park the tractor in the yard, shift it to neutral, set the throttle to idle, engage the power takeoff, and then dismount the tractor to lubricate the sickle. Working with great care (leaving any tractor while the PTO is engaged is a supreme no-no, something Dad drummed into our heads from earliest childhood), he held a pour tin at arm's length and drizzled used motor oil over the pentagonal sickle sections. At first the sections rasped and grated as they shifted with slow serpentine malevolence in and out of their rock guards; as the oil distributed itself, the rasp softened. When he had oiled the entire length of the sickle, Dad climbed back on the tractor and opened the throttle. Now the rasp of the dry sections disappeared altogether, changing phase to a deadly-sounding *snickety-snick*. The sunlight caught the sheened sickle sections and froze them, strobelike.

When conditions were right, the mower moved through the hayfields beautifully. The tall grass shuddered and danced on its stems for a split second after the sickle sheared it, then toppled backward in a continuous cascade to lie flat in the wake of the machine.

It didn't always work so smoothly. Sometimes wadded hay, sticks, or old cow pies blocked the cutter bar. Sometimes you hit a gopher mound or a small stone slipped through the rock guards and snapped a brittle sickle section. A good operator kept an eye continuously cast back for the telltale strip of unmown grass springing up like a cowlick through the fallen swath; the key was to notice it quickly, back up, and clear the blockage. If you left a green strip that went on for more than forty feet, you were in for some razzing.

By far the most maddening problem with the old sickle mower was the tendency of the whirling power takeoff shaft to snag the mown hay and in a split second spin up a bundle of hay so tight it cut the power to the mower and had to be hacked away with a jackknife. To counter this, Dad rigged a shield by suspending a plank on chains beneath the power takeoff. It worked pretty well. I have never in all my life heard my father curse, but years later when I was down beneath a serially malfunctioning hay-cutting machine in Wyoming, spittle-cussing and hacking away at the thirty-seventh impromptu round bale of the day, I wondered if just once in the gentle meadows of yesteryear that mower ever caused Dad to lose his religion.

By the time I was old enough to cut hay, the sickle mower had been relegated to pasture-clipping duty and Dad had purchased a New Holland haybine. The heart of the haybine was built around a sickle mechanism nearly identical to our old mower, but there the similarities ended. Mounted to the fore of the sickle was a seven-foot wide revolving reel fitted with spring-mounted steel teeth. The reel spun forward in the same direction as the wheels on the tractor but rotated at a rate exceeding ground speed so that the teeth could draw the hay toward the sickle, then—once it was cut—sweep it into a pair of rotating rollers functioning like a voracious wringer washer. Made of heavy rubber cast in mirror-image chevron patterns, the rollers spun at blurring speed. As the hay zipped through, the chevrons crimped the stems and bruised the leaves. This dramatically decreased the amount of time required to dry the hay, thus increasing our chances of beating the rain. As the hay shot from the rollers, adjustable fenders shaped the flow so it dropped in a clean-edged swath—much better than the old sickle mower. The entire machine was mounted on a wheeled frame raised and lowered by a hydraulic ram controlled from the tractor seat.

Whenever Dad sent me out to cut hay, he would assign a set number of "rounds." Because our haybine was the sort that would cut only in the wake of the right-hand side of the tractor, it was necessary to circle the field in a clockwise pattern, the perimeter of each "round" contracting by fourteen feet with every pass completed. I don't know that Dad had any formula for calculating the number of rounds, just that he was trying to strike a balance between having too much or too little hay on the ground at one time.

For a landlocked boy in northern Wisconsin, nothing substitutes for seafaring like nosing a Massey-Ferguson 132 tractor into an unmown hayfield on a sunny summer morning. The grasses part around the grille, rising as high as the engine shroud and sprinkling leafhoppers on your jeans. Rolling lobes of wind press across the meadow, made visible in shifting shades of silver as the seed heads dip and sway. Just inside the gate you pause for a moment like Columbus set to sail, discovery and depredation your call. And then you engage the power takeoff, roll the throttle back so the tach pegs around 1500 rpm, lower the haybine's clattering maw, ease out the clutch, and launch roaring into the uncharted grasses.

The first round was always the best. For better or worse, cutting hay appeals to the innate human need for control and order at the expense of natural things. The haybine goes gnashing into the organic tangle, and out the back comes a continuous straight-edged thatch that drapes the contour of the land like a woven green scarf, each round separated by a pale sun-starved strip of shorn stubble.

After the first couple of rounds gave me room to maneuver without smashing the standing hay, I reversed course and cut the outside round. We called this outermost pass "the backswath." Because it was up tight to the fencerows and woodlots, the backswath was often booby-trapped with fallen trees and dropped branches. Since the tractor was passing over this area during the first clockwise run, you had a chance to spot remove most obstacles, but invariably you missed a big branch that jammed the reel, or slipped through and into the rollers. The rollers were spring-loaded and designed to part and allow passage of a solid object, but because of the speed at which they were operating, *passage* isn't really the apt word—even the smallest solid object would cause them to slam open and shut with a *bang!* that cut through all the engine and machine noise and invariably bounced me half a foot off the tractor seat.

Once the backswath was flat, I returned to cutting clockwise. As with all fieldwork, you settle quickly into a groove. I can still summon my exact position on the Massey: left hand on the wheel, underbelly of my right forearm resting on the red fender, left knee against the gray-painted crankcase, right hand resting on the hydraulic control valve, upper body rotated slightly back and leaning right, head on a swivel. The position was a matter of function—you glanced forward now and then to correct your course and watch for corners coming up, but in the main your attention was directed to the machine behind you. You were watching to see that everything was spinning the way it ought to, that there were no strips of uncut grass popping up behind, and that the machine was taking as big a mouthful as possible— in fact, you steered mostly while looking backward at the position of the innermost sickle section, tweaking the wheel left and right to run as tight to the uncut hay as possible. Your left foot was always ready to cock and stuff the clutch if something went wrong, and you kept your right hand near the hydraulics control in case you needed to pop the header into the air to scale above a foxhole mound.

A morning spent cutting hay was a morning of being left to your own thoughts, with occasional nature breaks when the sandhill cranes came in, or a deer—bright rusty red against all the verdure—bounced across one corner of the field. But there were also plenty of reminders that you were running a monstrous machine through the habitat of hundreds of creatures: the constant rolling flicker of grasshoppers springing out ahead of the voracious reel; a skunk wobbling across the open swaths, headed for the cover of brush; a gutted gopher; a smashed mouse. Once when Dad had just purchased the haybine and was cutting the field behind the house, I rode my bike out only to see him stop the tractor, dismount, walk around behind the machine, and lift something up, up, and up until his arm was outstretched and the object was still nearly touching the ground. Squinting, I could see bits of furry color, and then I realized: chevron rubber rollers plus one barn cat equals one very skinny kitty carpet runner.

When the last round was mowed, I disengaged the power takeoff, pulled the trip cord releasing the pin that held the haybine in cutting position, and backed up at an angle so the hitch would fold back into road position. This placed the haybine more directly behind the tractor, making it easier to navigate gates and pass down the road without running two lanes wide. Once the spring-loaded slide pin popped into locked position, I raised the header until the hydraulics squealed, dismounted and went back to set the block designed to catch the header if the hydraulics failed. Then, back aboard, I pointed the tractor home. In the field behind me half the hay was lying flat in neat concentric squares, the first rounds already limp compared to the last, and in the center, the remaining hay stood sharp as a cut of sheet cake.

Back here in Fall Creek, out there in the beating sun, Amy is trudging through the grass as if she is being press-ganged into the Volga boatmen. Perhaps this is not the time to tell her haybine stories. I grab a second set of clippers and join her. We snip away together until her little red wagon is full with timothy, which we take to the asphalt in front of the garage and lay out to cure.

This is the time of year when the countryside truly thumbs its nose at the subzero purge of winter. The greenery is full-blown, the dew-drenched mornings reverberate with a tropical chirp and twitter, and everywhere there are babies: tiny rabbits beneath the apple tree, speckle-chested robins begging worms from mama, a spotted fawn by the mailbox down the driveway, and now and then a glimpse of the pheasant hen leading her loyal brood. From my desk I can hear the squeak of the swing as Amy bobs above the valley and the horizon beyond, and my heart is so light to hear this all through the open screens that I start singing along with the music I am playing, and so it is that Anneliese stops outside my office window and lets me finish a full chorus of Supertramp's "Give a Little Bit" before she politely knocks and enters, tickled to have caught me so unguarded and also I suppose to find me with an unknotted brow. "Oh, don't stop!" Anneliese says as I kick the volume down, and while the blush is still leaving my face she sits on my lap and with my arms around her, we talk like we haven't in a while. There is no question that I have bitten off more than I can chew this year, but there is no turning back now, and as we talk about how it's going, we have a chance in this five minutes to look each other in the eye and end by saying we love each other. Then the baby is crying on the monitor I keep out here on the bookshelf, but the thing that feels good as I watch my wife leave is that we are in this together. That night when I walk to the house after dark the entire valley right up to the yard is pulsing with fireflies.

If conditions were right, hay cut one day could be baled the next, but first it had to be raked. Rolling the hay with the rake flipped it off the moist ground where it had lain all night and fluffed it up so it might catch the breeze more easily; it also left the hay in a narrower strand that better fit the baler. We usually began raking by mid-morning, after the dew had come off. And we all loved to rake, because when you raked you got to drive the Johnny-Popper.

The equivalency is not absolute, but I'll pretty much guarantee you most farm kids remember their first moment at the wheel of a tractor with the approximate clarity of their first kiss. Me? Lisa Kettering, beneath a white pine in the moonlight on the road to Axehandle Lake, and Jerry Coubal's John Deere B through the gate beside the Norway pine with the pigtail twist. Nicknamed Johnny-Popper because of the distinctive two-cylinder *pop-pop-pop* of the exhaust, the tractor was a gangly-looking machine with tall rear wheels and a slim front end supported by two small wheels cambered to a narrow vee. The steering wheel was mounted in the near perpendicular and stood flat before your face like a clock on the wall. The square padded seat sat level with the top of the towering rear wheels, so you rode high, with a clear field of vision. Rather than a foot pedal, the B model had a hand clutch consisting of a slender steel rod capped with a round ball—rather like a solid iron walking stick. To engage the clutch you fed the walking stick forward; when you wanted to stop you pulled it backward, and the works disengaged with a steel-drum *ping!* Dad and his neighbor Jerry shared the Johnny-Popper back and forth during haying season. One morning when I was nine years old I went out back to watch Dad rake hay. When he was done, he unhitched the rake and let me ride back with him. On the return trip, we came to the gate beside the twisted Norway pine. Dad got down from the tractor to open the gate as he always did, only this time after he swung it open he looked up at me and said, "Why don't you take 'er through?" I still remember the offhand way he uttered the words, and how the adrenaline surged through me when I heard them. I realize now that he was probably anticipating my wide eyes.

The John Deere was a good starter tractor, because you didn't have to reach any pedals. The tall hand clutch, the position of the steering wheel, and a broad steel deck between the seat and the steering column made it possible to operate from a standing position—in fact when I was older I often drove standing up, if only because I could fantasize that rather than some hayfield in Sampson Township, I was navigating the Mississippi in a Mark Twain paddle wheeler.

Back there at that gate, with the John Deere going *pop . . . pop . . . pop* at low idle, I addressed the wheel with knees trembling. Reaching down to the gear selector, I ran it through its cast iron maze and into first. Then, with one hand on the steering wheel and heart tripping, I pushed that hand clutch slowly, slowly ahead until sure enough the green machine was inching forward, and there I was, *driving tractor.* The gate was plenty wide, but I felt like I was piloting the *Queen Mary* through a checkout lane at the IGA. When I passed through—head swiveling left, right, left to make sure I hadn't snapped the fence posts—I *pinged* the clutch out of gear with a combination of exhilaration and relief. Dad took the wheel back for the journey home, and I rode happily on his lap, still his small boy but much taller in my heart.

If you're going to train your youngster in tractor driving, hay raking is a pretty good first assignment. The rake is a relatively simple machine for a relatively simple task. Because it is ground driven, there is no power takeoff in which to become entangled, and when the the tractor stops, the moving parts stop. Also there is the advantage of turning the novice loose in a wide open field. Plenty of room for error, and if the kid gets drifty, odds are the worst you're gonna have is a windrow that wanders off course—as opposed, say, to a plow hooking forty feet of fence line, a cultivator ripping up half a row of corn, or a haybine trying to digest a pine tree. And because hayfields are dry by their nature, there is little risk of the kid freelancing and getting bogged in a mud hole. In short, it is tough to mis-rake hay. So for the nascent farmhand, a Johnny-Popper hooked to a hay rake is the equivalent of training wheels.

By the time I was old enough to saddle up, Dad had replaced our original rake (a rusty monster with oversize steel carriage wheels and a fixed hitch in front and wobbly trailing wheels to the rear) with a New Holland Model 256 fresh off the lot. Just like the old rake, it was a ground-driven side-delivery edition, but it ran on small rubber tires and was painted deep red and bright yellow. After pulling the rake into the field, I would stand on the hitch that joined the tractor to the rake and spin the plastic-handled cranks that raised and lowered either end of the reel— the key was to run the teeth low enough so that they combed up all the hay but not so low that they were gouging dirt, in which case you were alerted by little clots of sod smacking the back of your head.

When you got everything set right and got to rolling, the rake reel was a marvelous thing to watch as it spun counter to the direction of travel, the polished steel tine tips dancing in staccato flashes along the stubble line before swooping up and away around their oval orbit. Just ahead of the flickering blur the flat swath of dried hay rose and curled into itself like a wave angling for the beach, rolled over several times, then tumbled out to lie still in a fat unbroken rope. Sometimes a gust of wind would unroll the windrow and lay it flat again. Mostly you just went round and round and round. Jerry had mounted a suicide knob on the steering wheel, so when you got to a tight corner you could cramp that front end around right tight, then just turn loose and pull your chin back clear of the knob while the wheel spun back to straightaway. When I raked the back swath (we'd often wait an extra day or two as it was shaded and dried more slowly) I had to keep an eye out for tree limbs overhead because the John Deere rode so much higher than the Massey and the exhaust pipe stuck straight up in the air. (When the tractor wasn't in use, we capped the exhaust with a tin can—if the engine fired just right on start-up, the can would pop ten feet high.)

When the raking was done, it was home for lunch, and then the baling.

Still no chickens, but we've had the pigs for about a month. The passage of time has been marked by the daily evolution of the stunning subcutaneous rainbow chewed into my gluteus maximus by the frenzied coon dog. Lately the colors have moderated so it appears a thundercloud has parked on my butt. Like a remarkable version of Tom Sawyer's toe, this butt-bite is the sort of thing you just itch to show someone. I maintain my propriety, but have held the photographs in reserve and will make them available at auction should archivists of the proper caliber express interest and promise to keep everything high-tone.

Morning now breaks with the lids of the pig feeder banging. The racket reminds me the farm is alive, if only in that little corner. And it's nice to know they're down there fattening themselves up. Still, with each bang I realize it's a meter ticking on the feed bill, so we've been throwing everything we can at them food-wise. All of our table scraps, of course, but also green apples, dandelions, venison trimmings, and cleaned fish.

There is a profusion of wild grapes on our little farm. The vines wrap themselves around anything that stands still. The pigs are currently penned at the far end of an old overgrown paddock and concrete bunk feeder remaining from the days when this place was a going concern of a dairy farm. I'd like to expand the pigpen boundaries into the paddock later in the year, so I've been cutting back the grapevines a little each day. Having seen how they went for the nettles, I thought it worth a try to sling some of the vines in with the pigs. They went nuts, stripping the leaves off and chomping them down. So now every day I throw big armfuls in their pen and they snuffle right in there, ripping the leaves free and chomping happily, stopping only to fight with each other. The little female pig is forever nipping the boy pig on the ear and running him off from the best leaves and slop. I'd feel sorry for him except he's bigger than she is.

The only thing the pigs like better than grape leaves is pigweed (I grew up calling it lamb's quarters). It grows big and is easy to pull if the ground is moist, so it doesn't take long to collect a good bundle. They devour the stuff. We throw all our garden weeds in the pigpen. They snuffle through the quack and ignore the foxtail, but pouting their lower lips delicately, they worry out every last leaf of pigweed.

We got the idea to graze our pigs from reading Gene Logsdon's excellent *All Flesh Is Grass*. Gene makes the point that pigs were meant to grub and forage, and we're gonna test him on it. I do not know how he feels about feeding pigs grape leaves. In short, they will eat pretty much anything, but can be capriciously finicky. After weeks of gobbling every nettle I pitched across the fence, they've stopped cold. They give them a snuffle and move on. Perhaps the nettles reached a certain maturity and the taste changed, or they got too dry. It was always a marvel to watch them in the first place, with my legs stinging from wading amongst the nettles, only to see the pigs rummaging through them nose first.

I wander down there several times a day, often under the pretense of checking their feed, or to toss them a handful of dandelion leaves—on these they have never wavered, they fight over them—but mainly I just want to watch them. For all my talk about chickens being better than TV, the pigs hold their own. Already their personalities are emerging, and I find that Amy isn't the only one who will have to be reminded that they are not pets. But perhaps my concerns are ill-founded: some of our city cousins come to visit, and they all run down to look at the pigs. I follow them down and arrive to find Amy standing on the fourth rung of the panels, pointing first at one pig and then the other as she explains, "That one's Wilbur, and that one's Cocklebur . . . but in October, that one's ham, and that one's bacon!"

"Let's go check and see if the hay is dry," Dad would say after he kissed Mom and thanked her for lunch. Put your hay up too wet and it will overheat and you will wake up in the morning to find a giant smoldering briquette where your barn used to be. Reaching beneath the windrow down close to the earth where the hay was most likely to be moist, Dad would grab a shock of stems and twist it in his hands. If he felt the right lightness and crackle, the hay was cured and safe to bale.

We started on the hay wagons young. I remember standing beside the oldest Baalrud boy when I was still too small to do anything more than drag the bales from the chute to the back of the wagon. I watched how he stood sideways at the mouth of the chute, one hand hanging and the other resting on the emerging bale, and when it was my turn I stood just the same. Thus we accumulate the stances of manhood. I learned to ride the pitch and lurch of the wagon, knees slightly flexed to absorb the topography. I learned to wait until just before the bale reached the tipping point on the chute before hooking my fingers beneath the twin strands of twine, and how to walk with the bale to one side until I reached the stack and swung it round to boost it with my thigh. When I grew older and stronger I used a motion similar to a weight lifter's clean and jerk to get the hay bale above my head, where I would balance it on my forearms for a moment before bending at the knees and tossing it free-throw style atop the stack. My brothers and I marked our development as men by how high we could pile bales on a wagon. The day I pitched one nine high, I felt my shoulders broaden. Sometimes you'd rear back to pitch one and the twine would snap. The bale exploded in midair and dropped chaff on your head and down the neck of your shirt.

The hay wagon was towed behind a baler operating on a combination of forces ranging from the deft touch of the rake-like teeth that skimmed the hay from the stubble, the brute force of the knife-edged plunger chopping and stuffing the hay into the bale chamber, and the Rube Goldberg complexity of the knotter. In short, you fed a loose windrow in one end, and neatly bound bales came out the other. The speed of their delivery varied with the thickness of the hay—in thin cuttings the plunger had little to work with, and the bales moved in nearly invisible increments; if the windrow was the diameter of a grizzly bear, the bales lurched outward several inches at a time. A large flywheel kept the rhythm steady for the most part, but now and then—and especially on the backswath, where much of the hay lay in shade—you'd hear the baler bog on a chunk of wet clover, and then you'd keep an eye on the chute for the bright green slug packed between the paler dry hay. If there were two of you on the wagon it was fun to try to time out the bales so the other guy got the wet one, which would lift like a bag of bricks compared to the rest. After every few bales a soft pile of chaff would accumulate on the wagon below the up-tipped chute lip. I picked up the habit of kicking this pile away, but sometimes I would grab a pinch of the chaff and put it in my mouth like chew, drawing out the toasted sweetness of the dried alfalfa by squeezing it between my cheek and gum.

In between loads we dug the water jug out from where it was stored in the twine box beside the rolls of sisal that smelled of oil and Brazilian sun and unspooled from the center. We'd set the cooler on the edge of the empty wagon, unspin the plastic top and turn it over to catch the water from the miniature spigot, then pass it around. I remember raising the water to my lips and seeing bits of chaff skating the surface tension of the water. Our neighbor Jerry would always swirl water in the cap after the last drink, then sling the water to the ground before screwing the cap back on. A little ceremony before we went back to work.

As the day wore on and we circled ever tighter toward the middle of the field, mice would dart from windrow to windrow at the sound of the baler. By the time we were down to the last couple of rounds, they would pop out with regularity, and if one of the farm dogs was along we would leap from the wagon and sic the dog on the mouse. If there was no dog sometimes we just launched ourselves feetfirst and squashed the mouse with our boots. Often by the time we were passing back in the opposite direction, a hawk or crows were pulling at the carcass.

It was sometimes my duty to shuttle wagons back and forth across the field. If I got back early and the baler was on the far side of the field, I would turn off the tractor and lie beneath the wagon in the shade, and to this day I best remember haying as sounds from a distance—the up-and-down groan of the tractor engine as it lugged against the plunger, the delicate *clink-a-chunk* of the needles threading the knotter, the rumble of the power takeoff knuckles flexing on a tight turn. The days were vast and sunny, the school year was decisively over and the new one still unimaginable weeks off, and here we were in the country, putting up hay.

When the last windrow was consumed, we scrambled to the very tip-top of the load and rode home. The hay moved with a ponderous pitch and sway, as you imagined it might be to ride an elephant. It was quiet atop the bales, elevated above the tractor noise, and the ride home was relaxing. You could look out over the country. But the work was not done—the last of the hay still had to be stowed.

Unloading was the easiest job. You simply unpacked the pile and dropped the bales to the elevator, where hooks on the chain caught the bale and slid it up the rails and into the mow. The haymow was hot duty, and especially so if you were stacking in a steel shed. One summer we took a thermometer to the peak of the pole barn and it read 113 degrees. The person on the wagon was at the advantage over the mow crew, who had to carry bales across the uneven face of the stack. You were forever sticking your foot between two bales and going in up to your knee, the hay scratching along your shins. Every now and then the unloader would start dropping bales on the elevator faster and faster in a good-natured attempt to founder the folks in the mow. It was fun to see how long it took before a face popped out the haymow window and shot a dirty look.

When the last bale was stacked, I'd pull off my haying gloves. Dad bought them in stapled packs at Farm & Fleet. They were yellow and made from material something like felt. They were stiff the first time you drew them on but before long they went soft and balloony from all the sweat and the constant pull of the twine. If you wore a hole in a finger, the tip soon became packed with a solid knob of chaff. When you pulled your gloves off after a long day of haying in hot weather—especially if you were working the unventilated mow—your hands were wet and moist, almost dishpanny, and your wrists were matted with bracelets of sodden chaff where the cuffs had clung. It felt good, though, the cool air on your skin.

I take great satisfaction from watching the pigs strip nettles and eat grapevines, or churn through the quack in their pen, nibbling out the tender white shoots so that next year the soil has half a shot at growing something more useful. I like to think some of that chlorophyll is somehow working its way into the protein. One begins to understand the cachet of "grass-fed beef." Beyond the poetics, the stuff really does taste better, and regularly commands a premium price. That said, in the interest of stretching the food dollar, we happily feed the pigs whatever they'll devour, which is pretty much anything. When Anneliese heard that a local bread distributorship made its expired goods available for sale as animal feed, she called and got on the list. Basically you pay ten bucks and take whatever's on the shelves that day. I figured we'd get a few loaves, and that'd be good. Imagine my surprise when I walked into the back room and saw rack on rack. Now I'm pulling into the yard, and the rear of our thousand-dollar minivan (refusing to utter the m-word aloud, I call it the *fambulance*) is full from the floor to the windows. The variety is astounding: white bread, whole wheat, cinnamon raisin bagels, English muffins, hamburger buns, frosted cinnamon rolls, and bags of mini-doughnuts. I park the van in front of the garage, where it is visible from the kitchen window, and go into the house, looking for Amy.

I find her at her schoolwork. "Hey, snort-burger, when I went to town I bought some bread. I forgot to bring it in. Would you please do that?"

"Sure!" she says innocently. She is a sweet child, and therefore vulnerable.

"It's in the back of the van," I say. The minute she is out the door I wave Anneliese over to the window. "Watch this!"

The poor kid. Happily she trips up the sidewalk and across the drive. At the rear of the van she pulls the handle, and as the hatch rises to release the smell of yeast and reveal a stack of baked goods the size of a refrigerator, her jaw drops just as I hoped it might. For a full three seconds she just stands there gobsmacked. Then a bag of hot dog buns slithers off the pile and lands at her feet and she turns back toward the house, fists on her hips and a squinchy smile-frown on her face. I rush out to meet her, and by the time I get there she is laughing.

It will take us weeks to feed all of this and we don't want it to mold, so we jam as much of it as we can into our chest freezer, which is about half empty this time of the year. I cram it down (fascinated by store-bought bread after years of Mom's home-made, my brothers dubbed it "Kleenex bread" because you could take a whole loaf and scrunch it into a tiny wad), but even so we have quite a bit that won't fit. Anneliese keeps out several loaves of whole wheat and two bags of cinnamon-raisin bagels. Because they are technically expired, every single bag has been slashed with a razor—here in the land of overregulated plenty, people food becomes pig food at the stroke of midnight—but we trust our noses and are not picky. I make a mental note of which corner I stashed the doughnuts (those, I was careful not to crush) and liberate a tray of the cinnamon rolls.

Amy and I take two bags of bread down to the pigpen. They bite a few slices, then drift back disinterestedly to the wallow. When we come back later, most of the bread has been eaten, but a few slices remain. This is not typical piggishness. The following day we put the bread in a bucket, add water from the garden hose, and stir the whole works into a doughy mush. As soon as it splatters into the feeder they dive into it, smacking and snuffling and blowing bubbles, and in three minutes it is all gone. From then on, we always add the water.

I enjoy making trips to the feed mill in Fall Creek, and I enjoy lugging the bags down to the feeder, and I enjoy the sound of the feed slipping from the bag and the feel of the feed dust on my forearms, and when I replace the feeder cover and walk away I enjoy feeling that I have provided for my animals and that when they stick their snout in there, supper will be waiting. But all this doesn't come close to the feeling I get when I throw in a batch of grape leaves or a pail of scraps or a tray of expired cinnamon buns. *Free*, I think as the pigs gorge. *Free or cheap, and circling right back to the table.*

There is one limiting factor to the free-for-all buffet: my own queasiness. With the progression of summer, we have found ourselves overrun with cottontail rabbits. To quantify: when I step from the office on a recent warm evening, I count sixteen rabbits in the front yard alone. These are too many rabbits. Anneliese has been asking me to trim the herd for several weeks now, but I have resisted. I was raised never to shoot an animal unless the end result is bound for the table. And although I happily hunt and eat cottontails in winter, I was also raised to believe you should never eat a rabbit killed when the ground isn't frozen. Tularemia, the old-timers said. But when I saw those sixteen rabbits in one spot, I prepared to yield the point. Then a night later I went to fetch something from the pole barn and found a rabbit pulling itself weakly around the corner of the barn. It was obviously ill, all hunched up and blinking at me as I approached. I loaded the .22 and killed the poor miserable thing. When you have that many rabbits and they start showing up sick, it's time to cull. I went back up to the yard and shot the first rabbit I saw.

When I picked it up by the hind legs and walked to the weedy edge of the yard, I was just about to give it a fling when the old "shoot-it-you-eat-it" pang returned. I looked at it again. It was full-grown and to all appearances very healthy. Still, I couldn't shake the idea that you don't eat warm-weather rabbit. Then from down the ridge, I heard a querulous porcine grunt. Of course . . . Pigs are omnivorous. Rabbits are free. Waste not, want not. It seemed a little creepy, though. I waffled. Then I hiked on down there and slung that rabbit over the fence.

They snuffled at it a bit, and then the carnage began. They chomped it at opposite ends and ripped it in two. They crunched the bones. They gnawed the ears. They gobbled the guts.

Gentle reader, I am not a fellow quick to fold his tent in the face of grotesquery. But as I watched Cocklebur bounce the last bit of rabbit ear on her lower lip like she was dandling a cigar, the bridge of my nose assumed the topography of a crinkle-cut fry. I found myself wondering if tularemia could be passed on via pork chops, or if I was very possibly contributing to the spawn of mad porkrabbit disease.

I shot three more rabbits. I slung each one deep into the valley, where at night I hear the coyotes sing. There is more than one way to keep the circle unbroken.

While the project is on hold until definitive research can be conducted, I presume there is nothing inherently dangerous about converting our excess cottontails to bacon. The real challenge lies in coming up with a marketing program to make the idea as palatable as that of grass-fed beef. Those "grass-fed beef" people are working from a point of real advantage, as the term conjures bucolic images of breezy green meadows and trim cuts of pure protein. In days past I paid the rent by writing an advertising slogan or two, and I have applied myself to this current challenge with diligence, but so far have only come up with the undeniably catchy but ultimately unusable "Stick a fork in our rabbit-fed pork."

Like flowing water and snaking flames, the movement of hay—off the sickle, off the rake, into the baler—is hypnotic. And there are the aromatic dimensions—the hay green-sweet or minty at cutting, tealike in the mow. When I drive past a freshly mown hayfield I anticipate the fragrant seep and ride it right back to my seat on the Massey-Ferguson. When I step into my father's empty mow on a day when sunlight slants through the beams, the soft underbelly of my forearms tingles at the memory of the red dots and scratches left by the stem ends after a full day's baling.

And what better than haying to soothe the obsessive-compulsive beast? How clean the field looks when the last wagon departs. The stubble remains slanted in the direction of the last pass, and as on a checker-mowed lawn, you can read the bend of the stems and see how the day progressed. On tight corners the haybine always missed little bed-head tufts of hay. They bugged me like a collar sticking up, so sometimes I tried to trim them when I was done, but this plugged the sickle, so I'd have to shudder and drive home. But still: at the end of it all, you had the very green manifestation of summer swept cleanly from the field, pressed into cubes, and stowed in square corners against the winter. Every time I stack firewood, there is this moment at the finish when I step back and survey the neat row, and a yogalike calm fills me. It is the same with the hay pile. You look at it, and you think, Well, whatever the winter brings, we've got our hay up.

I spare Amy the bulk of my hayseed memories, but I do teach her to twist the timothy and listen for the crackle, to gauge the dryness against her palm. When it is ready, we pack it tightly into a cardboard box, and store the box on a shelf in the pump house, up off the ground so it doesn't reabsorb any moisture. The first day we fill the box maybe halfway. "There!" says Amy triumphantly. "If we're going to feed Guinea all winter, we're going to need at least three more boxes," I say. Her head and shoulders immediately droop. I call this *slumpage*. Slumpage drives me nuts, and as such I recently decreed that all slumpage would henceforth cease. So much for the dictatorship.

For all my talk of making hay and rites of passage, when my father calls and says he needs a hand getting the hay in this year,
it is Anneliese who packs the kids and drives north, leaving me
to write. Growing up in the valley across the way, she and her
sister used to work on the hay crew for Tom, the old farmer
down the road, so she can throw bales. We still visit Tom and his
wife now and then, and he's always got plenty of stories. Once
early on before Anneliese and I were married but headed that
way, Tom pulled me aside and told me Anneliese and her sister
had outworked most of the boys he ever hired. "One day they
told me they were tired of working with Stevie Wonder," Tom
said. "There wasn't anybody on the crew named Steve, so I said,
Who are you talking about? They pointed to this young fellow
who wasn't doing much." He was grinning now, anticipating the
punch line. "They told me, 'Every time he puts down a hay bale,
we Wonder if he'll ever pick up another one!'"

In late June we drive across the state for a family wedding and a
working vacation. The wedding reception is held in a beautifully
preserved old barn, and it's fun to watch my dad and brothers
standing at the edge of the dance floor scoping the timbers and
speculating on how the barn was constructed. The next morning
we drive up the Door County Peninsula and take a car ferry to
Washington Island, where I am to give a talk and perform my
first ever solo concert. I am nervous before the concert, as it is
the first time I will have ever appeared with just my guitar and

no one else to remember verses or play over my mistakes, but it's a passable show and I enjoy it once I relax. Coincidentally, the show is held in a converted barn. Our hosts put us up in a log cabin beside Lake Michigan. There are sandhill cranes on the lawn and goslings at the dock. Our friend Dan comes by with a pink Washington Island sweatshirt for Jane, and we visit until after dark. I have a chance to write in the Red Cup Coffee Shop, and we take Amy to see the smooth stones of Schoolhouse Beach. It is a good couple of days, but there is also that jolt of realizing how much world there is to drink in, and how much I miss when I get stuck in the vortex of my own just-in-time commitments. At the concert Amy takes a picture of Jane sleeping on Anneliese's shoulder as I sing. Later when I look at it I see Anneliese is smiling, her eyes bright, the way I remember them from the very first days of our courtship. Here lately with the baby and the insomnia but even more with my constant deadline-pushing, I have seen much less of that smile.

On the way back off the island, we are the only family on the ferry and the captain allows Amy to stand at his position in the wheelhouse. She puts her hands on the wheel, and when I see the size of her smile I can tell it is good for her to get the attention. When the ferry gate lowers, we drive our loyal van up the ramp to solid ground and point for home. Within the first mile, Jane starts her brass-lung bawling. It is 290 miles from the tip of Door County to our farm in Fall Creek. A six-hour drive. With the exception of twenty minutes somewhere west of Green Bay and the times we stopped to check to see if her diaper was wet, or if she was hungry, or if her seat buckles were pinching her, or if she was just worked up over the subprime mortgage mess, she screeched nonstop until we took her out of van and into the house at home. Honestly, I don't know how she did it without the aid of an oxygen tank. Clearly circular breathing was involved. Amy spent most of the ride with her hands locked over her ears. The next time someone suggests we put her in the car and drive around when she won't sleep, I will hand them the keys and the car seat and the ear muffs and tell them to have at it. Very early in the trip the air conditioner threw a belt, which was nice because it gave us reason to run with all the windows down, the siren sound of our bundle of joy pealing out across the countryside and scaring cows.

We've been holding off on getting the chicks from Billy and Margie until the Washington Island trip was out of the way, and now the big day has come. I spent the morning working with Mills, but since the coop is still nowhere near done, I used the time to build a chicken tractor. A chicken tractor is basically a portable enclosure with an open floor that allows the chickens to pick and scratch—you will want to read Joel Salatin for the definitive take, although this is the age when Google met chickens, so a profusion of examples are a split-second click away. Once they've worked the area over, you move the tractor, and they move with it. You can get as carried away as you like, but I kept mine simple. I built a wooden frame roughly the dimensions of a twin bed, incorporated a roost and a hinged door that doubled as a ramp, and even remembered to build a small shelf in the corner on which to place and secure the waterer. I put the whole works on two elevated skids and attached heavy rubber flaps along the bottom of either end. The skids allow it to be dragged from place to place, while the rubber flaps seal the gap at either end. The rubber is heavy enough to keep small chicks in and all but the most serious predators out, and because it can flap both ways, the tractor can be dragged from either end. Mills dug the rubber out of one of his Sanford and Son piles, smiling triumphantly. Nothing makes him happier than to put something he scavenged to good use. To finish off, I rigged hooks at either end of the skids so I could put loops in a rope and tow it either way.

I'm pleased with how it turned out. It's square and sturdy, and I managed to avoid idiot do-overs. In fact, with my sad record for building things, this went amazingly well, with only one head-knocking moment: because the pickup truck had been unavailable, I removed all the seats from the fambulance and took that. In a rare moment of prescience, I measured the inside width of the van's hatch and made sure to cut the chicken tractor cross-members a good inch narrower so it would fit in when it was done. Unfortunately, I overlooked the fact that the cumulative width of the finished product would include the vertical supports, and when I tried to put the finished product in the van, it was exactly two inches wider than the interior of the vehicle. I took this in relatively good humor, mostly because Mills was right there, and among handymen it is considered impolite to throw another man's tools.

But I was kinda stuck. I needed that tractor for the chickens I was picking up on the way home.

Miller looked at what I'd done, hooted, and said, "*SWEET MOTHER-OF-PEARL!*" Then he brightened. "Say! I've got a trailer! I got it at a *thrift sale*! I'm going to use it to haul wood behind my four-wheeler! It's right out back!"

Mills led me out behind the barn past several of his salvage piles and then pointed proudly at what appeared to be the remnants of a lawn-tractor accident in the weeds. Upon closer examination I identified the wreckage as a trailer because it had a hitch and two rubber tires, but the frame was bent, the plywood bed was delaminating, and the taillights were shattered. You could just envision the cop who pulled me over flipping his notebook open and getting comfortable before writing up all the violations embodied in this one little tangle.

But I didn't have much choice. We hooked the trailer to the van, strapped the chicken tractor to it, and I went on my way. Before I left, Mills and I came to an agreement on our story should I get stopped. I would tell the officer I had just purchased the trailer and was taking it home to make repairs. We actually rehearsed our story and the price—$85. You know, in case I got pulled and the officer decided to check my story. Contingency planning, you know. I like to think I respect the law enough not to feed them some silly half-baked story. I also strapped and re-strapped. I call this the "I-tried" strapping method. Yah, it's a tenfold rolling violation, Officer, *but I tried*.

I kept checking my rearview. In Mondovi I stopped at the hardware store to buy a chicken waterer and a feeder designed to be screwed on the bottom of a mason jar. Outside, noticing a couple of big-rig truckers checking their loads, I did a little circle around mine, snugging and twanging at the straps and checking the tires. I appeared pathetic and responsible.

Despite the scrap-yard trailer, the chicken tractor rode well, and soon I was at Billy's place. He was out back working on a chicken coop of his own. Several weeks ago Billy had bragged to me about his big score: just when he was trying to decide how to go about building a coop, someone whose kids had outgrown their backyard playhouse said he could have the structure. "It's *free!*" he said at the time. "All I gotta do is move it over here and drag it out back."

Well, yes. That was weeks ago. He decided to put it on a concrete pad. He decided to insulate. He decided to redo the roof. He decided he should repaint the siding. He and his wife painted the interior, then decided the color was chicken un-friendly and repainted it. When I get there today he is burying chicken wire in the dirt to prevent predators digging under. He is shirtless and pouring sweat. Billy is a big man and not well suited to heat. Since he got his "free" coop, he has made more trips to Menards than your average subdivision contrac-tor and has had at least one nasty incident involving tin snips. Billy is one of the gentler friends of my acquaintance, but when I find him out back today, sweaty and sticky beside the free coop that has now easily cleared four figures, he looks at me and hisses, "I am ready to stop *preparing* for chickens and just *watch* chickens!" I have abridged the quote, leaving out at least one contraband word.

He and Margie lead me to the garage, where the chicks—I'm not sure if you could still call them chicks; they are a month old, and mostly transitioned from fluff to feathers—are in the same plastic wading pool where I saw them with Amy, only they've grown and gotten more rambunctious and hard to keep in the pool. We transfer our dozen to a cardboard box lined with wood shavings. I load them in the back of the van and am on my way.

The trailer holds up fine, and I make it home without being arrested. Amy is visiting relatives and the baby is asleep, so it is just Anneliese who comes out for a look. It's late, so rather than try out the chicken tractor, we just transfer them to the old pump house (where I've rigged a temporary cage and roosts), give them feed and water, close the door, and leave them be for the night.

In the morning the fog is so thick I can hardly see the old granary across the yard. One by one I put the chickens into the tractor, then latch the trapdoor behind them. Leaning into the rope, I pull them onto a patch of green grass, and within seconds they are scratching and pecking like it's all they've ever done. Similar to the pigs—all their life in a plastic wading pool with wood shavings, and they know immediately what to do when put in contact with the earth. I lift some rocks and find a pair of angleworms. When I toss them in, the carnage is immediate. I go to the garden and pluck a couple of potato bugs—we've had a heavy infestation this year—but am disappointed when the chickens ignore them. We could use some potato-bug-eating chickens. Then one of the birds stretches, one leg and one wing back in the manner of a ballet dancer warming up before the barre, and I straighten and stand back just to watch for a while. It's dead calm here, the grass wet green, everything cottoned in stillness by the fog, nothing visible except a hazy semicircle of yard, half our house, and there in the middle of that yard, our chickens. When I return to the house, I meet Anneliese coming out.

We turn and stand together on the steps, looking at the scene. We have *chickens*. I move behind and put my arms around my wife, beautiful in one of my old flannel shirts.

Baby, I tell her, another dream has come true.

I am joking mostly, but standing there on our little patch with Anneliese in my arms, I hear the snap of the flag we are flying beside the driveway on this, the fourth day of July, and I think we have been blessed with a lovely little dream indeed.

All day I work in the office with a clear view to the chickens below. Up here in my swivel chair, I feel like a rancher of old, surveying my entire operation: two pigs, twelve chickens, one guinea pig (Amy puts him outside to graze). The fog burned off by midmorning, and now it's a fine day, peppered by the sound of a few early starters shooting off fireworks. Anneliese and I watched the chickens for a good while this morning, and I have gone down several times to move the tractor and feed them bread crumbs from our pig bakery stash. I throw in some windfall apples, but after a few tentative pecks they ignore them as they did the potato bugs. Later I will learn that if I slice the apples in half, they'll eat them quite well. When Anneliese brings Jane out and parks her stroller beside the chicken tractor, Jane regards the birds gravely and at great length, her chin tucked so that a second chin pops out. This is a mix-and-match batch—Black Australorp, Buff Orpington, White Rock, Speckled Sussex, Rhode Island Red, Partridge Rock, Barred Rock, Golden Laced Wyandotte—and they are relishing their relative freedom, every now and then exploding into short-lived but aggressive bursts of flight, and sometimes sprinting from one end of the pen to the other.

When evening comes with a storm threatening, I put the chicks back in the pump house. Amy's timothy is in there, boxed up and stacked for winter, and at the smell of it all my hay-making memories flood back. The sense of accomplishment when the hay was all baled, the wagons all emptied, the field all stubble. When I step outside into the lowering light, I remember how it felt to stack the last bale in the mow and to slide down the elevator rails, through the cool night air.

I hope I'm not working my poor daughter just to work her. I hold out hope that there are long-term benefits in assigning a child tasks that don't pay off with an immediate Dilly Bar. And while this life we are trying for here is a far cry from *real* farming, it does present opportunities for edification. This sort of thing can easily be overdone. I think of childhood friends who came to school only after several hours of choring, and how they were essentially unpaid help. And then there is the other lesson, the one that Anneliese is better at drawing out than I am—that not all tasks are completed on your own behalf. One late-summer evening I remember helping Dad and Mom push the year's last load of hay into the pole barn. The next day I would go back to school, and I remember how it was to stand there for a moment beside my parents, knowing that a winter's worth of forage was safely under roof, and that I had played a part in that. It wouldn't make me any cooler at school, but I had the sense that I had been an integral part of something worthwhile, something that paid an intangible dividend.

You won't find many hay-baling songs out there, but Fred Eaglesmith wrote a dandy called "Balin' Again," and there's a line in there about a man surveying his hayfields while having an imaginary conversation with his father. *Sure could use your advice on how to raise a couple kids*, he says, *I'm tryin' to raise 'em just the way you did.*

So I'm thinking of Fred as I watch my poor daughter again a week later, snipping more timothy and, yes, weeping. The things we do to the children.

Will it pay off?

I don't know. Looking at her there, I'm thinking maybe I'll write my own hay-making song, only call it "Wailin' Again." I am an imperfect father. This afternoon Anneliese asked if I could fold a batch of clothes before disappearing back into the office. I complied, but with slumpage.

CHAPTER 8

At some point every Sunday evening of my childhood there would come from the kitchen a steely rapping as Mom knocked a clot of Crisco off a soupspoon and into the popcorn pan. Like an albino slug, the white gob rode a self-perpetuating slick across the scarred pan bottom until it lodged at the low spot and puddled out. Sometimes Mom let me roll the spoon against the side of the heated pan. The residual Crisco clarified and ran from the widening hot spot until the spoon bowl shone clean, a modest but potent magic trick.

Mom kept the popcorn in a tin canister. Holding the canister against her body, she peeled back the plastic lid and—using a battered aluminum measuring spoon—dipped out a quarter-cup of kernels and poured them in the pot. They cascaded against the hot steel with the hiss of sleet pellets driven against a tin roof, sizzling electrically until Mom placed the lid, muting the spatter. Now and then we heard the abrasive scuff of the pan against the burner as she shook it to redistribute the kernels and oil. Eventually the first tentative pops came, and then a few more, and then like a metronome on a runaway came the frenetic firecracker rush like the whole string lit, miniature bull-snorts of steam escaping the lid until the expanding corn boosted it clear. Pressing the lid down with one hand and grabbing the handle with the other, Mom shook the pan again, coaxing a few more unspent kernels to blow. Then she dumped the contents into a stainless steel bowl big enough to bathe twin babies. The corn tumbled with a snowfall sound, an occasional old maid pinging the steel.

Then Mom gouged another knob of Crisco from the can and repeated the process. Between batches, she sliced apples and cheese, requisitioned one of us kids to move a stack of bowls to the table, and dumped sugar into the Kool-Aid pitcher. Whoever helped mix the Kool-Aid got to pick the flavor and lick the inside of the packet—a face-twisting treat that stained your tongue some fraudulent primary color. As ever, Mom was trying to do sixteen things at once, so the kitchen was often stratified with smoke from the inevitable burned batches. Anything short of cinders went in the bowl, and perhaps as a result to this day I fancy browned popcorn; the partial incineration imparts a malty nuttiness. When the stainless steel bowl was overflowing, Mom salted the whole works down, hollered, "Popcorn's ready!" and there was your supper.

The other delightful part of the Sunday night tradition was that everyone was allowed to bring a book to the table. The idea of being able to read while eating was delicious in every sense. My brother John read Jack London books and *Rascal* by Sterling North—you will not be surprised to learn he once fashioned a hat from the skin of a skunk and currently resides in a homemade log cabin. Dad usually read *Farm Journal* or *Successful Farming* or *The Agriculturalist*. The younger siblings brought picture books, and during his tenure Jud flipped through his omnipresent JC Penney Christmas catalog. I read my usual Tarzan and cowboy books, but I also remember holding *The Adventures of Huckleberry Finn* open with one hand while I shoveled popcorn into my face with the other. It was a broken-spined hardcover. There were illustrations within, so I wonder if it might have been an abridged version. Whenever I smell scorched Crisco, I think of Mark Twain.

It must have been a sight: eight to twelve of us packed around the dinner table, heads bowed over books splayed flat (somewhere a librarian cringes), the pages held open with one hand while the other dipped in and out of the corn, back and forth from bowl to mouth, the rhythm interrupted only when someone refilled a bowl or took a pull at their Kool-Aid. When your eyes are fixed on text, you tend to fish around with your free hand, and nearly every week someone upended their Kool-Aid. The minute the glass hit, Dad jumped up to make a dam with his hands in an attempt to keep the spill from leaking through the low spot in the table where the leaves met. For her part, Mom grabbed a spoon and scraped madly at the spreading slick, ladling the juice back in the glass one flat teaspoon at a time so it could be drunk. The same thing happened if someone spilled their milk. Sometimes when I wonder how my parents managed financially, I think of Mom going after those spoonfuls of Kool-Aid like an environmentalist trailing the *Exxon Valdez* with a soup ladle, and there's your answer.

Now that we kids have grown and have kids of our own, "pop-corn Sunday" has become the unofficial get-together night. There is no formal planning, you just drop in. Sometimes it's just a handful, sometimes the crowd is big enough that an additional table is required. Often it's brothers and sisters, but our friends and some of the neighbors also show up. For a decade after I moved back to New Auburn I lived six miles from my parents and rarely made it to popcorn Sunday. After I met Anneliese and introduced her to the tradition, she became the one who pushed for us to go more regularly. Now that we have moved farther away she is even more avid about keeping the date, and at least once a month she asks, "Are we planning on going to popcorn?" It makes me feel good, because I take it as a sign we have become quite solidly married.

Today the answer is yes, and Amy is tickled. She knows she'll likely see her cousin Sienna, and they will race toward each other on the sidewalk to hug with such aggression you fear they'll knock teeth loose. If her cousin Sidrock is there, she and Sienna will do their best to doll him up in clothes from Grandma's dress-up box, and then they'll all sit down at the play table and fight over who gets the green bowl and who gets the purple bowl.

We smell the popcorn as soon as we hit the porch, and when we step through the kitchen door it's a relatively full house. Mom and Dad and Tagg are there, and a little girl named Gloria Mom is car-ing for on a temporary basis. Gloria has severe epilepsy syndrome and is sitting strapped in her rolling chair beside the Monarch woodstove with her feeding tube hung from a hook on the wall. Mark and Kathleen are sitting on the piano bench, and Sidrock is charging around with a plastic dinosaur. John and Barbara are seated on the bench by the window beside Jed and his wife, Leanne. Amy and Sienna are already clacking around in high heels and tiaras, and just as we are dishing up the corn, our neighbors Roger and Debbie drop in. They have a truck farm down the road, and Roger and Jed share fieldwork. Roger is a John Deere man to the bone, and he sees to it that Jed's little boy Jake—currently roaring in and out of the kitchen with a plastic tractor—has plenty of green toys.

The table is the same as it always was, the Formica of the center leaf brighter than the rest because it sat in a closet out of the sun the first few years until the family grew large. Over on one side of the aluminum trim you can still see the saw marks from the remodeling days when Dad used the table as a sawhorse. When we were kids Dad sat at the head of the table, but tonight he's sitting on the oven door of the woodstove holding Tagg, who grins and drools per usual and waves the back of his hand at everyone who enters. Occasionally he pauses to woof or bite Dad on the arm. Mom sets the giant bowl of popcorn at the center of the table and Jed starts dishing up, the bowls passing around until everyone has one, the cheese and apple plate following, as well as a plate of vegetables. There is no Kool-Aid, but rather pop—the cheap stuff from IGA.

We don't read around the table anymore—too many grubbing little hands to manage that—but there is nonstop visiting. There is some discussion of current events and low-level nonmalignant gossip with careful circumventions around certain areas of politics, and a lot of stories from the past. Dad usually doesn't say much unless we convince him to get going. In my favorite moments someone will crack a good line and I'll look over and catch Dad with his head tipped down, his eyes closed, and his shoulders shaking silently. That's the full-on sign that you've caught his funny bone.

Jane is fussing, so I take her out through the porch and into the addition. Settling in the recliner, I cup her diapered butt in one palm and tuck her head beneath my chin, and shortly she is asleep. This is one of those moments I'm trying to soak in, to remember what it is for her to fit my chest like this. In the other room I can hear my family talking and laughing, and in here it's just me with the baby asleep and Jed's boy Jake grinning at me from beside the coal bucket that still holds the blocks Dad glued together with soybean paste all those years ago. Jake's favorite movie is *Cars*, and every now and then he says *"Pang!"* and tips his tractor back on its butt, just like in the film. Now Sidrock comes roaring in, and shortly after him Amy and Sienna, but Jane snoozes through it all, the noise of her generation drowning out the sound of the previous generation around the popcorn bowl in the other room.

The chickens are growing quickly, and scoot back and forth from the tractor to the pump house like old pros. At first we had to reach inside the tractor and fish them out one by one, and reverse the procedure in the morning when we moved them from the pen into the tractor. But now when I pull the tractor up to the pump house door and drop the gangplank, they skid down it on their heels, then hightail it straight into the roost. One poor little chicken is always last. She is wracked with constant tremors. They came on early and have persisted, so we call her Little Miss Shake-N-Bake. The tremors affect her gait, and it always takes her a few tries to hit the chicken tractor ramp straight on. But she's game. You can see her gather herself, resolutely struggle to point her wagging head at the door, and then, like the drunk choosing the one in the middle, dive for it. If she crashes into the side of the door, she simply gathers herself and tries it again. Sometimes I give Little Miss Shake-N-Bake a boost. As a result of watching her struggle, Amy has come to love Little Miss Shake-N-Bake as her favorite chicken.

The pigs are rapidly churning up their patch, upending clusters of quack, bumping up rocks, and now and then—based on the occasional barking seal noise that floats up from the pen— testing the limits of the electric fence. They are beginning to lose some of their charm, grunting aggressively and nipping at my calves when I enter the pen to refill their feeder. *Try lying down once and see what happens*, they seem to be saying. But it's still fun to grab the slop bucket and call out "PIG-PIG!" just to hear them woof and see them come bounding out from their excavations to press their snouts against the wire panels with their ears in the "What's up?" position. They quickly outgrew the rubber tub I bought at Farm & Fleet, so I have taken one of the many plastic barrels Mills scrounged from the dump and cut it in half lengthways to make a durable feeder. When I fill it with slop the pigs dive in feetfirst and fight for every morsel. Sometimes one of them slips and winds up sitting in the soup. Sometimes Cocklebur nips Wilbur in the ear until little spots of blood appear, but so far she has stopped short of devouring him, and he chows

on, smacking grotesquely and apparently unconcerned. Wilbur is
bigger, but Cocklebur runs the show.

I'm up in the office reviewing notes for a story, and Jane is
propped in the green chair again. She's still a tiny little bean,
and I can still balance her on my forearm, but she's fattening up
some, getting a little marshmallowy in the legs, and rounder in
the face. Lately she's been working on holding her head steady.
She hasn't quite got her cranial gyro dialed in, and there is a lot
of bob and weave. Sometimes she'll really get to wobbling, and
you can't help but think of Little Miss Shake-N-Bake. She's also
been working hard to summon her first laugh. We'll make faces
and her eyes will crinkle and her mouth will twitch up, but then
she just sputters and gacks and hacks. The other night when Anneliese was bathing her in the bathtub, I leaned over and asked
Jane if she was happy and I swear she said "*Uh-huh!*" but then it
was back to happy drooling and there has been nothing since.

Here in the office now her face has begun to crumple. I switch
the music from Tom T. Hall to Gnarls Barkley and turn it up. Her
head bobbles in the direction of the speakers, and I have bought
myself three more minutes.

Down beside the pigpen, the sweet corn is tasseling. I weeded
it just once, and it came on remarkably well. The soybeans, on
the other hand, have been all but swamped by the quack. It is
midmorning, and Amy and I have come down to feed the pigs
zucchini. If you just chuck whole zucchini in there, they tend
to ignore it, but we've found that if you chunk it up they'll have
a go at it. Instead of using knives or even a spade, we slam the

zucchini against the wire panels. If you do it hard enough, they dice themselves. It's a very satisfying transformation. Cockle-bur seems to be lagging behind Wilbur size wise. She's nice and healthy looking, just smaller. I haven't wormed the pigs, figuring that since they're the first pigs on this patch in twenty years if not forever, it's not necessary. Now I'm second-guessing myself, so I wait until Cocklebur goes over into the bathroom corner (pigs tend to defecate in one corner of the pen only) to do her busi-ness, and then I crawl over the panel and study the poop, kicking it apart with the toe of my boot. I don't see any worms. Maybe she's just smaller because she's a girl.

I send Amy up to the house to turn on the hose tap, and we fill the wallow. This has become our favorite activity of the day. The pigs revel in the water, sticking their snoots into the stream, closing their eyes, and letting the water play over their cheeks and face. Sometimes they chomp at the water, and oftentimes Cocklebur gets so worked up she stampedes herself in tight stiff-legged circles, her chunky body teeter-tottering fore and aft. Then she flops at the rim of the wallow and slowly rolls until she goes over center and slides right in. When the sun is hot and the pigs are caked with dried mud, something about the water hit-ting their skin gives them the itches. They lean hard against the shelter and rub back and forth. Sometimes they back up to a steel post and wag their hindquarters back and forth to hit the right spot. Today I lean in with the grass whip and scrape it back and forth across each pig. The dust flies off their bristly hides, and they grunt happily. With Wilbur, if you hit just the right spot he groans and his knees give out.

On the way back to the house I notice the hose connection is leaking. The brass fitting is squashed to an oval. Apparently it got run over. I'm making a trip to town later today—I add a hose repair kit to the shopping list. Amy and I make sandwiches and eat them on the deck. We've had a great morning of being pals. After lunch, I grab a shovel and we go out to the compost pile to dig angleworms. When we've got a nice couple of handfuls, we take them over and drop them in the chicken tractor just to watch the chickens fight over them. In between bites of worm, they clean their beaks by swiping them back and forth in the grass. Next we raid the pigs' bakery stash for a bag of English muffins and scrub them across the poultry wire. This has a cheese grater effect. The crumbs shower down, and the chickens peck crazily.

Like the legendary bullet unheard, the worst bad news rarely gives warning, but rather drops on your head without so much as a shadow to announce it. Think of a feed bag filled with lead shot and allowed to achieve terminal velocity before the dead thump, the impact so echoless and mundane that for one dumb moment we fail to recognize the devastation for what it is.

When the cell phone rang the first time, I did not hear it because I was wandering around the home improvement store and had left the phone in the cluttered front seat of my car. When it rang the second time, I had just departed the lot and was merging to the frontage road. I fished it out, flipped it open, and put it to my ear. Anneliese, her voice dreadfully calm: "You need to go to the hospital. Jake had an accident. He's coming on the chopper."

Over twenty years now I have responded to emergencies in the instant, and for the first time ever I went dead blank. I remember the car moving silently down the road while I tried to put the name in context (*Jake . . . Jake?*), tried to get a fix on the message, tried to know what to do, and then the familiar cold focus cleared my brain. I flipped the turn signal and turned right opposite of the way I had been headed.

My brother Jed knew he wanted to be a farmer from the time he was in diapers. He was still in them when he climbed aboard the Ferguson tractor and managed to punch the starter and get it lurching forward, although thankfully the key was off so he didn't go far. When he was a preschooler I built him a haymow on the doghouse and rigged a pulley system so he could pull up the miniature hay bales that Dad made for him by hand-tripping the baler knotter. When he got older, my folks had a tough time getting Jed to maintain decent grades in school—not because he lacked the aptitude, but because he simply didn't see the need for any lessons not available on the farm. In high school they homeschooled him for a while, mainly to prevent him languishing in a classroom. Mom taught him to make bread, how to cook and can, and how to patch his own jeans. Dad assigned him farm-based math problems and lined him up with a work-study job at the feed mill in Chetek. He pushed railcars, unloaded feed, and learned some agribusiness.

He returned to school for his senior year because he wanted to graduate with his friends, and then as soon as the cap and gown were stowed, he began muscling out a living. By turns a farmer, a logger, an over-the-road trucker, and a laborer for any occasion, he supports his family doing whatever it takes as long as it's honest and borderline legal. For a while he did custom choring—milking cows and overseeing the operation for farmers who wanted to take a week off, or needed temporary help. He was good at it. A couple of farmers came back to find milk production had actually gone up in their absence. Word got out, and he got hired a lot. In the meantime, he was carving out his own living—running a joint milking operation with a friend for a while, doing custom fieldwork, and always the logging and truck driving. He saved up and bought the neighboring farm and took to raising crops and young stock. He got some pigs. And after years of bachelorhood, he found a blond country girl named Sarah and married her. They had been married seven weeks when Sarah was killed in a car accident. Jed answered the call as a member of the fire department and was the first on scene. He did as he was trained, but it wasn't enough.

The darkness was unimaginable. But he emerged, and married Leanne. She came to the marriage with Sienna, a beautiful three-year-old girl. It was good to see the new little family at popcorn Sunday nights, with Jed smiling again. In time, a baby came—Jake. Jakey, we called him, or sometimes Jaker. Jed will tell you that first year wasn't easy. That he second-guessed the whole idea of babies. But by that second year Jake was toddling, and he became Jed's constant companion and mimic. When he picked up even the lightest object he grunted comically, like Daddy did. When he took a swig off his bottle, he followed it with a breathy, overacted *"Aaahhhh!"* just the way Daddy taught him. In my favorite photo of the boy, he is in the back of Jed's truck, surrounded by chain saws, hard hats, a plastic tub of bar oil, scattered wrenches, and Jed's firefighting gear. His diaper is low-slung and dirt-scuffed, and his little hands are grease-lined as any mechanic's. He is hatching a grin like he has come to know the whole wide world, and in the shadowed background Jed is standing with the driver's door open, looking back at his boy, holding him steady in his gaze. I shot the photo on one of those cheapo disposable cameras. Somewhere along the line the camera wound up under the seat of my car, and it was a year or so before I found it again, covered in lint and fluorescent orange Cheetos crumbs. We shot up the rest of the roll and sent it in, and when the pictures came back, there was Jakey, only by that time Jakey was gone.

Jakey died, and there is no poetry in it. When Jed's wife died, I asked his permission before writing about it. This time I can't even bring myself to broach the subject. The night is scalded on our souls, and I am not going to tell much. Everyone tried so hard, beginning with Jed, who pulled Jakey from the farm pond just moments after the boy disappeared. Praying for a heartbeat, Jed and Leanne worked together to revive him. They are both members of the local fire department, and later they would say their training just kicked in. Soon they heard sirens, help coming the way it always has in rural settings—from friends and neighbors suddenly turned rescuers. Then the ambulance came, and then the chopper, and when it lifted away with Jakey inside, the fire chief put Jed and Leanne in their vehicle and drove them the forty-five miles to the hospital. I was waiting outside the emergency room when they arrived, and what I will remember forever is Leanne running to be with her little boy and the solid feel of my brother's muscles even as he sagged in my arms.

Everyone worked so hard, and we were in the little room with Jake for a long, long time. We knew there was little chance, and at the end there was none. In the hallway I saw firefighters, paramedics, nurses, the emergency room physician—everyone in tears. Mom and Dad had been traveling toward the middle of the state when they got the call, so we all gathered in the open air of the parking lot until they arrived, and Jed and Leanne got in the back of their car and everyone went home.

While I was driving to the hospital, Anneliese had called our neighbor Ginny to come and sit with Amy and Jane. Both of our vehicles were running on empty, so coming home from the hospital we stopped for gas. While the pump ran I was standing beside the car feeling the absolute weariness grief brings, and when I looked up across the fuel island to Anneliese, our eyes met and I saw the very same weariness in her. There was something in that moment—on the concrete under the false light, the anonymous cars coming and going all around but our eyes wordlessly speaking—that reminded me why I love her and how. In her weariness I saw compassion.

When we got back home the children were asleep and Ginny was at the kitchen table. We told her Jake was gone, thanked her, and she left quietly. Her husband Ed, the man who tilled our pig patch, had recently been diagnosed with cancer. She knows grief of her own.

Upstairs, we looked in on Amy, wrapped in her sheets. And then I went to the crib and bent down, listening close in the dark until I heard the silken thread of breath, in and out, in and out.

I wept then, my wife beside me.

In the morning we pull the chicken tractor out, fill the feeders, slop the hogs. I move the tractor a little too quickly and Little Miss Shake-N-Bake gets rolled out the back, squeezed between the cross-member and the ground, and then swept along by the rubber skirt for a few feet. I figure I have ruined her for good, but when Amy runs to pick her up the bird evades her for the first three passes, a bona fide sign of life.

Even without me running her over, the bird's tremors have gotten worse. Amy picks her up several times a day, smoothing the feathers along her back and cooing in her little chicken ears. Little Miss Shake-N-Bake is a determined bird. Naturally she is always on the outside fighting her way in when it comes to dinnertime, and you can't help but root for her, lowering her head to slam into the wall of tail feathers before her and then bouncing back like the skinny kid hitting a blocking sled at football practice. She ricochets, shakes her head like a woozy prizefighter, and charges forward again. Her single-mindedness serves her well, if she's going to survive as the runt; when we throw table scraps into the tractor, the other birds tussle with each other and dart from scrap to scrap, seemingly more intent on coveting than eating. Meanwhile, Little Miss Shake-N-Bake gets herself a chunk of cucumber and just sticks with it. It takes her a while to get her beak dialed in, she shoots wide a lot, but she is indefatigable, and even though she bats about .250, the cucumber slowly disappears.

Up north on the home farm the phone will be ringing steady. Cars and trucks will be coming in the yard. Our place feels quiet and removed. There is the urge to just drop everything and head to my folks', but one thing we have learned is how friends and neighbors come in and fill these early days. I spoke to Mom earlier, told her to call if there is anything we can do, and I know she will, but in the meantime, there is life to be taken care of. Amy has swimming lessons, and after that, piano lessons. We go. We do. What else? In the chicken tractor Little Miss Shake-N-Bake has cornered another piece of cucumber, and when the other birds come after her she dives beak-first beneath the corner shelf that supports the waterer. Only her tail sticks out as resolutely she digs in.

On visitation day, I drive to Fall Creek to buy pig feed, and then I drive into Eau Claire to buy dress shoes for Jed, as he has none of his own. He called yesterday and asked if he might borrow mine, and I said yes, but then I got to looking at them. I bought them some years back in a fit of stylishness and they are square-toed verging on floppy—imagine a cross between a clown shoe and a pilgrim shoe. I won't put him through that. Instead I buy him the plainest sort of black shoe I could find. Then I go back home and feed the pigs.

Then we pack up the family and drive north.

It is a long, long day. We stand together just beside the casket and the line goes right out the door for hours. They arrive steadily—relatives, neighbors, distant cousins flown in, fire department members in uniform, church people I haven't seen since some Sunday meeting years ago, and many faces I just plain don't recognize. There are a lot of old farmers who can't bear to look in the casket, and you see these sunburned old dogs approach my brother and break down weeping as they take his hand or wrap him in their bearish arms, and maybe they are wearing big belt buckles or unmodish jeans or have their sparse hair Brylcreemed in the style of a '60s trucker, but it strikes me again how much we miss if we rely wholly on poets to parse the tender center of the human heart. At times like this I am grateful I was not raised to be sleek. Behind us pictures of Jakey project on the wall, dissolving one into another, and beside the casket are all his green tractors from Debbie and Roger, his John Deere blanket, and the wooden biplane his Uncle John made for him by hand, because if it was possible Jakey loved anything more than a green tractor, it was airplanes. *"Oh!"* he'd say at the first sound of an engine overhead. *"Whassat?"* And then he'd stand stock-still, watching until the engine faded and the plane was gone. He came by that innately, because he certainly wasn't pointed to it by the ground-bound farmers who raised him. At one point John slips away, and I see him kneeling before the toys, carefully tipping the tractors over backward, one by one, until every single one is sitting nose in the air. Leanne remains at the casket, stroking Jakey's hair and greeting the mourners one by one. How thin and pale she looks, and yet she will not sit or turn away. A tall man leaves the line, approaching her with tears in his eyes, and I recognize him as her fire instructor. Just over two years ago he marveled when Leanne showed up to complete her firefighting class in full turnout gear, exactly one week after Jake was born.

We stand there, brothers and sisters by blood and otherwise—Suzanne has come, and Don and Migena, and Kathleen. Donna and her husband Grant have come to care for Jane, allowing Anneliese to be by my side. I reach for her hand much of the day. And of course there are young ones everywhere, clambering in the pews, running in and out, hollering happily as they play tag beneath the churchyard swings.

That night Jed and I are on his lawn, talking quiet in a pair of canvas chairs, leaned way back to watch the sky all thick with stars. Now and then a big jet passes above us, so far up as to be silent. When Jed worked late in the shop across the yard, Jake would hang out with him, dragging big wrenches across the concrete, riding his plastic tractor in circles, and just generally getting grubby. He'd stay happy at it so long Jed says time would get away and when they walked to the house it was dark and Jake would want to stop and say good night to the stars. He'd pick out those blinking lights, Jed said as we watched another silent airliner slide across the sky. You can imagine the two of them then, faces to the heavens, the little boy with his finger extended, tracing a light seven miles high.

"I was wondering if you could rewrite this," Jed says, digging a folded and refolded piece of paper from his jeans. "Kinda smooth it up." It's the eulogy. "I want to try to read it," he says. "Prob'ly won't make it, but I wanna try." I pocket it, tell him I will do.

We talk past midnight. Jakey was a little roughneck, and not at all retiring. But whenever they looked at the stars, Jed says, the boy spoke in a whisper. He'd point to the moon, Jed said, look up at me, and whisper, *cookie*.

It is a short walk down the road to my father's farm in the dark, and beneath the stars I think of Jed and Jake hushed there in the yard, and I wonder, what does a child sense, that he would address the universe in a whisper?

I let myself in quietly, but my parents are both in their recliners downstairs, Dad dozing fitfully and Mom reading her Bible. I power up Dad's computer and unfold the eulogy. It is written in ballpoint, in a scraggly but readable hand, and the more I read, the more I realize there is little for me to do. I retype it anyway, stopping to bawl between lines, but in the end I alter maybe four words.

There is humor—the story about Roger teaching Jake that *one end of the cigarette is hot!*, and how Jake made chain-saw noises when he cut his food. The line about Jake and Jed spending their time either working, goofing off, or goofing off working. I recognize my brothers in that. There is more, but it is not mine to share. When I reach the part where he tells about Jakey whispering to the stars I bawl again, and knowing Jed will never make it through, print an extra copy for the minister. And finally I climb the stairs to bed, to one of my childhood bedrooms, and stare straight up in the dark. I am remembering that before Jane was born, I was talking to a friend about how it was when he went from one child to two. "Love expands," he said, "to fit the need." I am wondering if grief can do the same.

Jed reads the eulogy straight through. When he nears the part about the stars, tears are streaming down my face because I know what is coming, but he takes it absolutely and resolutely home. Then there is the terrible closing of the casket, and we leave the church. At the cemetery little Sidrock says loudly, Jakey drowned and now they are burying him, and you can feel the collective instinctive move to say *shush!* but then the ebb on the heels of it as we know it is a time when the truth should be left as it is. When the service is complete and we prepare to leave, John steps to the casket and draws flowers from the bouquet, handing them to the children as they file by. I carry the vision of his fingers, thick and grease-lined, passing the slender stems one by one to the tender hands reaching up, toward the sun beating in the sky.

Our meat chickens have arrived. They came in the mail, peeping in their perforated box. In order to hit the price break, we're splitting a batch with our neighbor Terry. With the coop still not done, I have built a small crate that I have placed in the garage atop an old piece of linoleum to keep the concrete clean. With the pump house already occupied by the layers, the garage is the only space available that we can seal up tight and varmint-proof. I rig a heat lamp, and first thing the next morning I discover the drawback to my plan: the things smell *awful*. I leave the door open during the day, but by day two the smell has already penetrated the cement blocks. Out of kindness Anneliese has not inquired, but I have told friends she has every right to ask: Which came first, the chicken or the coop?

We've begun to free-range the layers, dispensing with the tractor and just turning them loose. They love the new freedom. They run and swoop. They flutter and hop. They get all chesty and pushy and face each other down in pecking matches, neck feathers flared into a fright wig muff. They scratch and chase flies and dandelion fluff. They did this in the chicken tractor too, but within a few hours everything was tramped down. I'd pull the tractor ahead ten feet and they'd be happy again, but soon everything would be flattened and I'd have to skid the whole works again. Now they have the wide world at their disposal, and they can't wait to tear it up. Two of the hens discover an anthill and scratch at it with great exaggerated motions like they're going for a major peelout. When they unearth a scatter of white eggs from the hill, their beaks jackhammer the earth like zigzag sewing machines. When I fed them apples in the tractor I had to slice the apples up to get them started, and even then they'd peck at them just off and on. Now when they range under the apple tree they drive their beaks deep into the wormholes and peck fresh white craters into the apple meal. Poor little Shake-N-Bake lags behind, wobbling as she does and having to sometimes come to a full stop before gathering herself and plunging forward again, but eventually she winds up under the apple tree, and just as I've seen her do with a cucumber before, once she picks an apple she stays with it, hanging in there even after the other chickens have charged off after grasshoppers. She's smaller than the rest, no doubt due to the fact that it's harder for her to eat.

When she does chase off after the other chickens she'll get four or five good strides in and then go into a tumbling veer, like someone reached out and pushed her from the side. The thing that really gets you when it's a fun group sprint across the yard is how fully she believes that she can run with the crowd; she never gets hangdog or stops, she just collects herself and goes bounding off as if this time she'll be the smoothest bird in Eau Claire County. And it isn't always veering. Occasionally one of her legs will straighten explosively and she'll shoot a foot and a half into the air. It really is something to see. Sometimes when she's doing her dangedest to hang with the pack her earnest ping-ponging schizophrenic hopscotch is so over the top I find myself laughing out loud. I mean no disrespect, but honestly it's like watching a fast-forward version of the classic Tim Conway bit where he plays a dentist who shoots one leg full of Novocain.

In perhaps the saddest funny moment so far, I am throwing bread on the lawn when poor Little Miss Shake-N-Bake gets overexcited, rears back to take a hefty stab at a crumb, misses, and jabs her bill in the dirt. She literally has to back up and tug it free. As penance for laughing right out loud I give her an entire slice of cinnamon raisin bread—I figure she can aim at the raisins with the rest of the slice as backstop.

When we left the church after Jake's funeral we carried a big box of cards, and they've been coming in the mail daily ever since, so on a Sunday we all gather at the farm to write thank-yous. Barbara with her tax accountant sense of organization is in charge, and she has us ranged around the kitchen table in stations to handle everything from slitting the envelopes to noting the contents to writing and stamping the notes in return. As each card makes its way through the system, we read it, often checking the return address to place the person. There are a lot of *Oh!*s as we recognize old familiar signatures, while other times it takes a group effort to match the name on the return address with a person. For all we would give if this day could be taken away, the net effect of all these envelopes circling is that the afternoon slides into a sustained conversation in consideration of all that is good in this world, and when the last stamp is affixed, Mom pulls out the pan and starts popping corn.

Three days later I am back at Jed's, ripping out fence line. A while back when he heard I was scavenging steel posts, Jed told me he was going to be reconfiguring the field north of his house, and I could have the posts if I'd help pull them. When he called last night to see if I could help today, we both knew it wasn't about the posts. I've got to be in Madison—four hours south of here—by evening for purposes of researching yet another writing assignment, but I knew I had to do this first. We're working in the deep grass at the border of the field. The hot weather has been holding pretty much unabated—90 degrees again today, with humidity to match. The sweat is running off the bill of my cap and my shirt is soaked through and streaked with rust from the posts, which have been in the ground for decades. We're working smart for once. Jed is in his skid steer, and I have a chain. I sling the chain around the post, give it a couple of wraps, and then hold tension on one end while hooking the other over the skid steer bucket. You can crank and yank on posts like this all day long and get nothing for your trouble but a sprung back, but the skid steer plucks them from the earth as easy as a straw from a malt. We move quickly from post to post, all along the edge of the forty. As soon as they clear the earth, I unwrap the chain and chunk them in the bucket. We're working right out by the road, and at one point two of our longtime neighbors—Big Ed (who used to work at the feed mill) and Gerald— pull over on the shoulder and we visit. Big Ed asks me about my pigs, and I tell him about our stash of bakery bread. "Oh, that's the best thing to feed pigs," he says. "Bread and withit."

"Withit?"

"Yah," says Big Ed, his eyes twinkling. "Bread and whatever comes withit!"

The talk is light, with no mention of the trouble, but when Jed lost Sarah, these two men were always showing up just now and then at the right time, and when they drive off it makes me feel better knowing they'll be circling in the long days ahead. We go back to pulling posts. At one point we have to work around a telephone pole, and when I give it half a wrap of chain and raise a quizzical eyebrow in Jed's direction, he tilts his head and grins. It is the blessing of dumb work done close to the earth—one gritty minute at a time, we move forward.

When the last post is in the bucket, Jed says there are a few stacked out behind the shop, down where Big Mama the giant pig is living out her retirement. He'd rather not go down there, he says. The first time he fed the pig after Jake died, he found Jake's little plastic grain scoop in the dirt. He figures that's where the boy went, down to give the pig some feed like he loved to do, only this time he wandered on. I go down there myself then, and while I'm digging the posts from the weeds I'm thinking how for Jed and Leanne everything in sight has become a dreadful connotation. Months after my sister Rya died, Dad went to the basement for firewood, and looking up at the old defunct ductwork he broke into tears, remembering how Rya used to sit beside the heat register upstairs and they would call back and forth to each other.

When I get back out to the yard, the pastor has arrived. All my usual reluctances are in place, but I have been watching this man, and he is doing good work. I shake his hand and leave grateful, knowing my brother is about to sit down and take counsel. As I drive away, I turn on the radio and learn the stock market has fallen 300 points, and very clearly I think, *Whatever.* The drive to Madison is long, and the hotel room when I get there seems a cube of unreality.

It is a blasting hot day when I return home—the dried clay around the hog wallow is bleached white in the sun—and the dang pigs have destroyed their only source of shade. The hutch I put together several months ago is ripped to bits, flat as the secondhand particleboard I used to build it.

I'm not sure why they chose today. Perhaps they were just bored, or perhaps overnight one of them gained the quarter pound necessary to collapse the wall during the afternoon butt-scratching session. I was working up in the yard when I heard the rending sound of the tarp being torn in two. When I got down there, the wall I had wired to the steel posts was still standing, but the other had collapsed. My first reaction was, Hey, I'm surprised it lasted this long. It was hardly built to code, and that's what I get for roofing it with a blue plastic tarp. And pigs by nature root and push and bull against everything. It was bound to happen. But my equanimity got a little thin as I drew nearer the pen and realized: rather than running off to some neutral corner and staring back like some kid who swears it wasn't him who broke the sugar bowl, the pigs were actively—no, *joyfully*—finishing the job. Cocklebur is gnawing on a section of two-by-four. As I approach, she takes it in her jaws and, with a toss of her head, flips it across the pen like a puppy flinging a chew toy. Wilbur is snuffling around the one collapsed wall, looking to find purchase for the rim of his snout. When he finally hooks it beneath a section of particleboard, he bulldozes forward to the sounds of more tarp tearing and screws being stripped from the wood. Wilbur circles around again and pokes his nose through the tear in the tarp, then his entire head. He stands there blinking for a minute, then plunges his body forward, the tarp ripping and popping loose from its staples. Not wanting to be left out, Cocklebur sprints around in a tight half circle and low-hurdles through the hole Wilbur has left.

Worried they'll cut themselves on exposed screws and also hoping to salvage as much of the material as I can, I climb in the pen and start trying to chuck remnants over the fence, but this only seems to excite the pigs more. They're on a full-bore happy rampage now, gallivanting and woofing excitedly, standing on boards I'm trying to lift, gnawing on the tarp, and generally wreaking happy havoc. Every time I try to pick up a board, they run over and put their front hooves on it, or take bites at the wood so close I can feel their breath and get slobber on my fingers. Frustrated and not interested in feeding my digits to pigs, I ball up my fist and smack Cocklebur right on her wet snoot and she gives out a high-pitched grunt and jumps back a foot, but then comes boring right back in. By the time I get the last shred of the shelter thrown over the fence I'm tickled rather than upset. They are absolutely single-minded in their dedication to destroying what I had built, but they are just so playful about it all. Absolutely vandalous creatures, but gleeful in their depredations.

I have another tarp in the shed, so I grab four bungee cords and suspend it above one corner of the pen so they've at least got shade, the knuckleheads. Even as I'm walking away Cocklebur is standing tippy-hoofed with her snout in the air, trying hard as she can to get a bite of the new tarp, but she is built far too low. The tarp is safe, and the fun is done.

Jane and I are back in the office. She has had a fine nap, and Anneliese has taken advantage of the time to make a grocery run. Jane sucks her thumb and beams at the ceiling, which is nothing but white texture. I get down before her and we talk some. At first she can't be troubled to unplug her thumb—she keeps her forefinger hooked over the bridge of her nose—but then she decides to talk, and her brow furrows and her gaze grows earnest, and she works her lips, but after all that it's still just gack and hack. Then she starts bicycling and making spinach faces, which means the storm is gathering and the squarelip is not far behind. Hearing the van, I gather her up. Let's go help Mom unload groceries, I say, and then I wonder when exactly it was I began calling my wife Mom.

The day we buried Jake the funeral procession was winding through the country to the cemetery when a biplane appeared in the sky. High enough that it looked like a gorgeous yellow toy, but low enough that you could see the shine and polish of the fuselage, and the blue star painted on the underside of each wing. The plane was moving right to left, and crossed the road directly above the fire truck driving point. After proceeding a gracious distance, it rose slightly and banked a slow turn, then flattened out to cross again, this time left to right. And so it went for the next ten minutes and eight miles, the line of cars moving sedately down the road, the biplane tacking gracefully windward and lee. When we arrived at the cemetery the craft rose to circle in the distance. The engine noise receded to altitude.

We were walking to the back of the black Suburban containing Jake's casket when Jed squinted at the sky and nodded toward the plane. "What's the plan?" he asked me. "Not sure," I said. The biplane is owned by a friend of ours. John had given him a call. Jed looked square at me, and for a split second I saw the old reckless flash.

"Well, I hope he gives 'er hell."

We drew out Jakey's little casket and bore him to the grave.

To the best of my recollection it has always been sunny when our family has convened at this tiny place. I don't read the sunniness as any sort of sign, just note it. It was sunny when we buried my sister Rya after her heart and lungs finally failed her little soldier spirit at the age of six. I was a junior in high school then, bound in a few short weeks for a cattle ranch in Wyoming. It was sunny when we buried Eric, just ten years old and nine years older than the doctors predicted—which is not to fault the doctors, as they failed to factor in my mother. I was at loose ends in those days, out of college but trying to find my way. And it was sunny when we buried Sarah, just feet from where Jed is standing now, facing the only death possibly worse.

The pastor gathered us in close. There would be a prayer, after which we would linger, leave the cemetery slowly, the children each with a flower. But first the pastor drew our attention back to the airplane, which had descended again and was approaching from the south on a line parallel to the cemetery fence. When the airplane drew even with us, well above the treetops and some two hundred yards to the east, the nose lifted and it began to climb a quarter circle until it was pointed straight up and then it continued on around, until it was upside down, the wheels at zenith. As the plane broke over to complete the loop, the engine stalled and went silent and remained so for a breathless pair of seconds, and then black smoke puffed from the cowling and we shortly heard the cough as the engine fired and caught and the craft carved a slow turn back to the south, nose pitched to take on altitude. There were smiles then, even laughter. Little dressed-up cousins pointing. *A loop-the-loop.* That was good, we thought. Imagine Jakey watching that.

The biplane shrunk in the sky then, rising lazily away. Again the engine noise faded and we turned back to the grave, prepared to pray. But back in the distance the motor modulated up a half-pitch. We turned to look, and the sun flashed from the left wingtip as the opposite wingtip dipped, and now the plane was curling back toward us, dropping swiftly. The downward arc steepened to a plummet, and looked precipitous to the point of danger—surely he was falling too fast—but still the plane descended, drifting sideways until it was approaching over the farm buildings to the south and still dropping, now nearly straight at us, and just when you thought *No, too low*, the wings fixed themselves dead square to the earth and now the noise came on flat and furious, the plane over the corn tasseltop high and distorted behind a heat mirage, and the roar grew and grew and the plane blasted through the shimmer to bellow toward us terribly vivid now, flat-out thunder on a rope, and when it was nearly upon us a gloved fist shot from the cockpit in a rock-solid salute, and in that split second the plane twisted steeply up and left and up and left, the fist still high, and then the plane just rising up, and up, and silently up, and then nothing and with it our hearts into the white-hot sky.

CHAPTER 9

One summer evening when the other kids got to go swimming, I had to stay home in bed. There had been some infraction. I no longer recall the offense, but I can summon with absolute clarity the sand-crackle sound of car tires departing the driveway, the soft swell of acceleration, and the fade to distance. Staring at the ceiling from beneath one thin blanket, I felt starkly alone as the sun lowered and I imagined my siblings boisterously en route to Fish Lake, their beach towels slung brightly across the seats.

Tonight Amy is living her own version of my past. The evening before Anneliese and I were married, we held a yard dance. A string band called Duck for the Oyster provided the music, and their caller Karen led us through the quadrilles and contras with such verve and simplicity that even an arrhythmic clomper such as I had a delightful time. We have attended several of their events since, and Amy especially loves them. This weekend they played up north, and it was our plan to attend as a family.

This did not come to pass.

A sweet girl, our Amy, but as with any developing child, there are low-level intransigencies, the cumulative effect being that the dictatorship must intervene. In the matter of gathering hay for the guinea pig, there was slumpage unabated; piano practice had become a weeping sit-down strike interspersed with spates of enervated tinkling; spelling lessons began to feel as if they were being conducted in a room stripped of everything but a chair and one naked lightbulb. Sensing that Anneliese was nearing the end of her rope (I pick up on this sort of thing, especially if she writes a note and tapes it to the toilet seat), I intervened with a series of expostulatory disquisitions blending themes of personal responsibility, the virtues of alacrity, respect for one's elders, the long-term benefits of good posture at the piano bench, and a general review of all-American gumption. Once I actually *harrumphed*. I truly believed I was getting somewhere until—just as I was hitting my stride on the delayed gratification of hard work and a job well done—Amy looked up at me through her tears, stamped her foot, and howled, "But I only want to do the *FUN* stuff!"

I found her logic impeccable and wished I could cut to commercial.

Despite my one-man Chautauqua act, there was no improvement. Anneliese and I talked and agreed it was time to implement measurable standards backed by that euphemistic woodshed, *consequences*. A family meeting followed, the chore and school list was reviewed, and standards of performance were clearly set. We were not as forthcoming about the consequences, as it has been our experience that specific carrots generate short-term bounces evanescent as a last-minute campaign promise. What

Amy couldn't know was that a Duck for the Oyster dance was the prize behind Door One. If she didn't hit the mark, she'd be staying home with me.

The critical morning dawned with hope. You root for the kid, you know. How quickly as parents we discover that it really *does* hurt us more than them, and I dreaded the evening if she failed. By mid-afternoon it was clear she would fall short. Even with a gentle reminder here and there, she kept dawdling. When the time came and Anneliese began wordlessly packing Jane's diaper bag, Amy sensed that something was up. "Where are we going?" she asked. "I'm going to a dance," said Anneliese. "Oh!" said Amy quickly. "I'd better get my chores done, then!" Of course there was nowhere near enough time, and we broke the bad news.

"But what am I going to do?"

"You have to stay home," said Anneliese quietly.

A flood of tears. And then Amy wailed, "You mean I have to stay home with *grumpy old Mike?*"

Anneliese had dinner with my family at the farm that night. She says my brothers couldn't decide what tickled them more— the fact that Amy called me grumpy or old.

To see the realization set in, to see her sweet hopeful face crumple, to hear the tears that followed as Anneliese drove away . . . ach, it rips the heart out of me. I leaned backward against the sink as she wept and wept at the kitchen table, and it felt like I had kicked a bunny. How many years before we knew what good or damage we had done? For a while I just let her roll, then I announced that it was time to eat. She kept breaking down as we got supper ready, but I slogged on. There were brief moments of lucid conversation interspersed with extended crying jags. As we ate, the ratio slowly reversed itself, but by the time the meal was over I was shot, and proposed we just put the chickens away and head for bed. As we snuggled in for the bedtime books, Amy said, "Tell me the story about when you couldn't go swimming."

I had forgotten that I had told her the story previously. As part of one of my sermons, no doubt. And so I told it again, and we talked about why parents do what they do, and then I read her a book about a girl who loved the color pink and then I kissed her good night and in the morning it was another day.

Today when I turn out the layers, the Speckled Sussex and the Barred Rock are slow to move. Rather than scooting away when I reach for them, they allow themselves to be caught. I can't see anything visibly wrong with either one, so I just leave them in the pump house and go to the office. By mid-afternoon I notice that the Barred Rock has eased her way outside. She's tentative and doesn't rejoin the flock right away, but she's clearly improved. The Speckled Sussex is right where I left her, motionless except when she blinks. I place a saucer of water right beside her and she dips her beak twice, but even that movement is desultory and shortly the other chickens swoop in and stomp all over the saucer. I have no idea what the problem is, and decide to treat it with a dose of wait-and-see.

In the evening Anneliese's mother babysits while Anneliese and I go out for our third anniversary. For the past two years we have celebrated in a small cabin beside Lake Superior. This year my schedule won't allow it and the pigs and chickens make it harder to leave. We have a nice meal and then go for coffee in a strip mall. Every anniversary we review our vows, and as we go through them tonight, it isn't the shoreside discussion in the pines overlooking Gitchigume with the waves breaking below, but at least we are face-to-face, talking about something other than diapers and chickens. (I have lately developed a persistent habit of steering all conversation toward the topics of coop ventilation, the effects of molt on the laying cycle, and personal poultry anecdotes. In honor of our love, Anneliese has placed a firm one-night moratorium on chicken stories.) This year as in the two years previous, the session splits pretty evenly between reminiscence (each line shakes loose happy snippets of memory) and the equivalent of a polite but firm visit mediated by an auditor representing the Department of Weights and Measures. It's bracing to see your promises there in black and white: "I will treat you with reverence . . ."

I wrote the word *reverence* into our vows in honor of the way my father has always treated my mother. Dad taught me that reverence wasn't fawning, nor was it always delivered in hushed tones. I saw it in the goofy way he doffed his fur-lined Boris Yeltsin hat when he opened the van door for her on Sunday mornings; the way he quietly abstained when we kids teased her for not getting our jokes; the way he never failed to leave the dinner table without thanking her. And there was the reverence *between* them: lest we be deceived, on many occasions—together and separately—Mom and Dad made sure we understood that their marriage had rough patches and disagreements, but that they had long ago promised to work it out quietly behind closed doors. It didn't hurt that they sometimes made sure to let us catch them kissing. Nothing off-putting, just a hug and peck in the kitchen or in the sheep barn during lambing. In this I believe they were extending their reverence to the children—letting us know that when we went to sleep it was in a house headed by parents joined at hip and heart.

Tonight in the strip mall as we revisit the other words we promised each other that day (*gratitude . . . devotion . . . trust . . . unity . . .*), my eye is continually drawn back to *reverence*, and how the animation of the word requires more than simple respect or careful talk. I am thinking reverence requires presence and attention, and that I must bestow reverence on my wife if I wish it to fall gently on my children. Looking up from the vows between us, I see a delicate brown fleck set against the blue of Anneliese's right eye. I discovered the fleck the first time Anneliese allowed me in close, but haven't noticed it in some time. I need to look my wife in the eye more often.

We hold hands on the drive home, and while Anneliese goes to the house I close the chickens in the pump house. They are mostly roosted and fluffed. As she always does, Little Miss Shake-N-Bake has settled in the wood chips on the floor. The struggle to roost is a challenge beyond her at the end of day. The Speckled Sussex is exactly where I left her earlier. I refresh her saucer of water, turn out the light, close the door, and drop the hook in the eye. Then go in the house and to bed, and begin the fourth year of my marriage.

Mid-afternoon of the next day I look up from the desk in time to see the Speckled Sussex step tentatively out of the pump house. She continually cants her head to the side and shakes it like a swimmer with water in the ear. I'm sure some poultry expert could diagnose this. I just stare at her. She steps carefully, and when she pecks at the grass she is tentative, but it seems a good sign that she's up and about.

And so it is disappointing when I open the pump house door the following morning and there she is flat on the floor, stiff as a board, dead as a nail. Well, shoot, I think. Picking the feathered corpse up by its feet, I walk down past the burn barrel and sling it deep into the ravine. Fox food. Unless the coyotes find her first, and they probably will. Ever since we began free-ranging them, I've been compulsive about counting the chickens whenever I see them. I adjust the tally in my head, take it down from a dozen to eleven.

On a humid overcast morning three weeks after the death of his son, I meet Jed in Chippewa Falls. My stepmother-in-law is letting me salvage her old pigpen, and Jed is bringing his trailer to haul the panels, which are too long to fit safely in my truck. The old pen is back in the brush and weeds, so everything is woven in the overgrowth. It takes a lot of ripping and tugging to get the panels loose, and the steel T-posts are even harder to free. We're in the middle of a month-long drought and the rock-hard dirt holds them like concrete. I am a complete doughboy in comparison to Jed, but we do have enough shared raising that we know how to hit the traces in unison. At one point we're reefing on a panel, trying to lever it free with a length of pipe, when the whole works collapses, smashing my little finger and raising a walnut-sized lump on my forearm. "That hurt?" he asks, chuckling gleefully. "Pretty much," I say, smiling back.

We load the panels and head on down to Fall Creek. I have

some dead trees that need felling, and Jed has brought his logging gear. I have a chain saw, but a couple of these trees are monsters. In my twenty years making ambulance calls I've found more than one squashed corpse whose last act on earth was to sink a saw blade into a tree trunk. Jed has been logging every winter for years (and is furthermore a graduate of logger safety school), so it makes sense to ask for his help and stay out of the way. It takes him less than two hours to fell, limb, and section up the trees; the same task would have taken me at least two days. While he logs, I run the tractor back and forth, dragging away limbs and pulling the larger sections out into a nearby field where I can cut them into firewood lengths later.

When we're done and Jed has thrown his gear in the truck, Anneliese comes out. We talk about how he and Leanne are doing, knowing full well there's no sufficient answer. We are talking for much the same reason we have been working together this morning: there are things that have to be done, and also we are finding reasons—quite literally minute by minute—to keep moving. The word *closure* is tissue paper over a tar pit. In these early days the best you can do is find ways to stop screaming while your psyche begins the sand-grain trickle of sorting the nightmare.

So we talk some. And then we say *seeya* like we always say *seeya*—no lingering, no look of meaningful intent, just *seeya*. As the truck and trailer rumble out of sight around a bend in the driveway, Anneliese and I hold hands and ache for the pain we cannot absorb on behalf of those we love.

I use the salvaged posts and panels to expand the pigpen. The pigs tear into the new sod, their tails spinning. I am watching

Cocklebur snout through crab grass roots when she pauses, dipping her head up and down. When I look in closer I see she has teased an angleworm free from the dirt and is feeding it backward into her mouth with flicks of her almost prehensile lower lip.

The soybeans didn't survive the weeds, but the sweet corn is thriving. Each day I cut and feed the pigs several stalks. They eat the cobs and chew the leaves. We have also come into another cheap pig-food bonus—our friends Kenneth and Virginia Smote have an excess of goat milk, and they have been saving it for us. Once a week we bring it home in buckets, and each day I mix it with the expired baked goods. Kenneth claims he has raised fine pork on goat milk alone, and I have promised him some pork chops in time. We don't have enough refrigerator space to store all the buckets, and by the end of the week I am decanting some diabolically clotty fondue, but those pigs slurp it right down. On the downside the buckets don't seal well and we have a bumpy driveway; I have noticed that on real warm days the inside of our van smells of curdled goat.

We have a number of apple trees on the property, and when the first windfalls dropped I happily gathered buckets of them for the pigs, but I have been frustrated. The first time I tipped them into the feeder, the sound of them tumbling against the plastic brought the pigs a-bounding. How disappointed I was when after a few nibbles they wandered disinterestedly away. I couldn't bear the idea of free pig food going to waste, so for a few days I took to making worms-and-all apple smoothies in the blender and stirring them in with the bread and goat milk. The smoothie technique worked, but after several days of cleaning the blender,

I settled for just lobbing a few over the fence now and then. I did come past the pen one afternoon to find Wilbur gazing at me with a big red apple in his mouth, and I admit I imagined him on a sterling silver platter and said "Hold that pose" out loud.

The pigs are getting big. Jed told me I'd know they were ready to butcher when they wouldn't fit between my legs at a straddle, but (A) Jed's about five-foot-five—if he followed his own advice he'd never butcher a pig bigger than a schnauzer, and (B) you gotta be kidding. Tom, the old-timer in the valley, says four pounds of grain equals one pound of gain, but I haven't been keeping track, plus there's all the goat milk and expired cinnamon rolls. He also says any pig over 250 pounds is starting to run more to fat than meat. My brother-in-law Mark says I should just raise them as big as I can get them. As he put it, "I figure every additional inch is two more pork chops."

I ask around, and someone says there is a fellow who will come out to the farm and do the butchering. His name is Muzzy. I give him a call, ask if he'll come and eyeball the pigs, let me know if they're getting close.

My sister Kathleen and brother-in-law Mark are butchering their chickens, and I've gone to lend a hand. Mark once helped me resurrect my beloved old International pickup, so I owe him pretty much forever, and furthermore, I'm treating this as a refresher course in case we decide to butcher our own chickens this year. I haven't butchered chickens since I helped my brother John about six years ago. Mark and I wade into the chickens and each grab a bird. The things are huge and solid—I feel like I'm cradling a feathered bowling ball. Mark has rolled up four tin funnels and nailed them to a horizontal plank attached to the back of his chicken tractor. We stick the birds headfirst down the funnels. When all four funnels are occupied, Mark grabs a butcher knife, fishes out the head of the first chicken, extends the neck and with one quick motion, severs the head, and moves to the next bird. The funnels keep the birds from flapping all over creation (while the idea of a chicken running around with its head cut off tends to be used in a humorous sense, the reality is much more unnerving, and more than one farm kid recalls the freakout of zigzagging across the yard two steps ahead of a spasming two-legged gore-geyser that seems to be matching zig for zag) while also allowing the blood to drain to the ground.

When the blood is down to a slow drip and the legs stop kicking, I grab two birds by their feet, pull them from the funnel, and carry them over to where my father is heating a large pan of water over a propane burner for the purpose of scalding the birds, which loosens their feathers for plucking. The scalding is tricky—immerse the bird too long or at too high a temperature, and you begin to cook the skin and it will tear when you pluck; too briefly or at too low a temperature and the feathers fail to loosen, and plucking becomes even more of a chore. As to the perfect temperature for scalding, experts agree: the experts disagree. Dad finds it helps if you plunge the bird up and down slowly, which seems to increase the penetration of the hot water.

Next Dad gives the bird a turn on the automatic plucker. My brother John built the plucker himself and its main feature is a rotating drum bristling with rubber fingers. Dad lays the chicken on the drum and turns it this way and that. The fingers snag the feathers and sends them flying into a pile. As the day progresses they collect on the lawn like a soggy pink-tinted snowbank. The plucker gets about 90 percent of the feathers (lately we have been eyeballing an improved model), but the rest require manual labor. At eye level between two trees, Mark has rigged a line threaded with a series of blunt hooks. The hooks are sized so that the shin of the chicken will easily fit but the foot catches. Most of the remaining feathers can be plucked by hand, although we do break out the pliers now and then. At the very end we use a handheld propane torch to singe the pinfeathers, which blacken, curl, and burn away to nothing. If you hold the flame in one place too long, the skin begins to contract and you are quite literally cooking chicken. Next Mark and the neighbor lady cut the feet loose at the joint, eviscerate the birds, and put them in tubs of water to cool. A hose is placed in the tub and allowed to run at a trickle so the water stays cold and refreshes itself, rinsing away what is officially known as *skack*.

Once we're under way, by about the sixth bird or so, we establish an informal division of labor and a rhythm sets in. I stick mostly to plucking and killing. I do not like the killing part, and find the best thing is just to move decisively. There is the usual unavoidable nonmetaphorical instant of recognizing exactly what it takes to enjoy chicken dinner, but I resist the temptation to deconstruct the process further. It's miserably hot, which highlights the delight of working elbow deep in guts and wet feathers. Wasps continually alight on the chicken carcasses and buzz at our ears when we shoo them. Sidrock has picked a severed head from the pile beneath the funnels, and, squatted beside a tree, he is working the beak open and closed and poking at eyeballs. Jed has arrived, and before he starts plucking, he grabs a chicken foot and exposes the white ribbons of tendon with his jackknife. Then he shows Sidrock how to make the chicken claw open and close by tugging on the tendons. Sidrock is openmouthed in wonder. "Go show your mom," says Jed slyly, and the little boy tears off for the house with the claw in one hand. Flopped in the shade beneath the four-wheeler, Mark's dog is hot-mouthing a rooster head. The deep red comb has gone yellowish pale, and when the dog settles in to gnaw it from the skull, the sound reminds me of the one I hear in my own head when I am chewing gristle.

Jed joins in, working and jesting with the rest of us. But there are new lines around his eyes, and after an hour he puts down his knife, lays back across the ATV seat, puts his cap over his eyes, and sleeps. No one says anything, but we know we are seeing the absolute weariness of grief.

Forty-three chickens go through our ragtag assembly line. I am not staying for supper, but I wish I could because on the way to the car I walk past the charcoal grill and catch the scent of beer-can chicken, which is made by roasting a chicken in the vertical position with an open can of beer stuck up its hinder. A final indignity, I suppose. The whiff I catch on my way to the car makes my tummy grumble, and the stinky feathers clinging to my boots do nothing to diminish my appetite.

As for our meat chickens, they are growing at an alarming pace. Within two days of delivery they were sprouting wing feathers and already they are approaching Cornish hen dimensions. It's tough to love the meat chickens. They stomp around thick-legged and flat-footed, and when I turn them out on fresh grass or give them sweet corn on the cob, they peck some, but mostly they

just sit and wait for ground feed. The free bread they ignore entirely. They are nearly impossible to move in the chicken tractor, lollygagging confusedly. Rather than startle forward when
the back of the tractor bumps them in the butt, they often flop
over and rest quietly while it rolls over them. We gave them
chicken starter to begin with, but now that they're coming on,
we've switched them to hog feed, since it's cheaper than chicken
feed. The one time they show life is when I replenish the feeders,
at which point they trample and ram each other without mercy.
Once for the sake of my own entertainment I filled a mason jar
with feed, capped it, and set it in the pen just to watch them
peck madly at the glass. They are clearly bred solely to generate
protein, and in the first couple of weeks I had to build three different temporary boxes for them, increasing the size each time.
Once we started free-ranging the layers I switched the meats
over to the chicken tractor, but still, every night I have to drag
them into the garage, as the chicken tractor isn't strong enough to
withstand more aggressive predators, and just the other day I saw
a fisher (basically a weasel on steroids) cross the driveway. When
I'm gone, Anneliese has to drag them back and forth. The garage
is deeply pungent, and every time I go in there I am reminded of
my shortcomings. None of this would be an issue if the chicken
coop was done.

The laying hens, on the other hand, are great fun. They follow
me to the office in the morning and tap on the glass of the storm
door. If I rattle a tin tray of feed, they come running. I catch and
feed them grasshoppers. They are not pets as such (their siblings,
the ones that remained with our friends Billy and Margie, actually jump into your arms for cuddle time), but they are engaged

and surprising and fun to watch. Sometimes when the deadlines are really closing in I wander down and let them eat feed from the palm of my hand just for the relaxation of it. I can envision a time when all a man would want is a porch, unoccupied time, and chickens in the yard.

Jane is not growing at the rate of a meat chicken, but she is holding her own. When I put her in the football hold my arm gets tired before she does. I've begun taking her along in a backpack while I'm doing chores. Yesterday when we approached the pigpen and the pigs did their usual woofing and galumphing, I heard a funny noise behind me, kind of a slobbery chortle, and when I craned my head I could see Jane smiling at the pigs and I realized I had just heard her very first out-loud laugh.

Nighttime, however, has not been so joyous. After settling into a groove in which she slept most the night, suddenly Jane's begun waking up and bawling. Last night after Anneliese had tried feeding her and she was crying again, I took my turn. I went through my whole repertoire of tricks—rocking, bouncing, pacing around the kitchen island sixteen times in the ambient glow of the microwave light—and nothing worked. Finally I gave her my knuckle to suck, and as she latched on I felt a slight snag and there was your answer: her first tooth breaking through.

I was down tending the meat chickens when Muzzy the butcher rolled into the yard driving a shiny red diesel truck with a winch and boom mounted in the bed. *Custom Butchering & Scrap*

Iron, it says on the driver's side door. He seemed to step from the cab while the truck was still in motion, already in stride as his feet hit the driveway. I was a ways across the yard, but I could see he was long-legged and cowboy-slim. His bill cap rode high and looked big for his head. The brim wasn't tracking with the rest of him—it pointed leftward. He was wearing a silver pistol in a black leather holster. The pistol rode loose, with the handle tipped out, and I noticed he had no fingers on his left hand.

He was stomping toward me, but he was looking down toward the pigs. The female was visible beside the feeder. "Oh, she's nice!" he said. "That's a good one! From here I'd say about two-twenty!" I shook his hand—the one with fingers—and walked him down to the pen. While walking I shot a glance at the hand and could see it had been patched up with flaps and grafts. Wherever those fingers went, they didn't go easy.

"Oh, look at that big guy!" he said, pointing to the male, who had come snuffling up from the back of the enclosure. "He's nice and thick through the shoulders. He'd be ready now. He'll go about two-fif . . . no . . . I'd say two-sixty." He had clambered over the gate and was right in there now, hands spanning the hog's front shoulders, poking and squeezing some. "Dandy!" he said. "That's great, thick through the shoulders like that." He was looking up at me and smiling, proud as if he'd raised 'em for me. Naturally it felt good to hear that the guy liked our pigs.

"The female, she'll go about two-thirty," he said, revising his original estimate upward. "Yah, you could butcher 'em anytime. Give my wife a call and she'll set it up."

We had been sitting Fritz the Dog again, but that is over now because he killed four of the laying hens. It took him a matter of minutes. When I walked to my office the chickens were beneath the big pine tree and Fritz was lying beside the sidewalk, and when I came back a few moments later there were feathers in the yard and Fritz was hiding behind the pump house. The White Rock was dead beside the light pole. Nothing was left of the Partridge Rock save a few brown speckled feathers. The third chicken had for all intents and purposes evaporated. But worst of all, there beside a pinecone in the grass was a segment of wing that I recognized immediately as a remnant of Little Miss Shake-N-Bake.

When Fritz tore up the cold frame, I flat lost it. When he killed the chickens, I felt something colder. I immediately flashed to the day I got off the school bus and found Dad stringing up dead sheep on the corncrib. Several of the sheep were horribly wounded, the flesh gnawed from their back legs, gaping chunks torn from their hams. Snags of wool hung loose from their bellies. I remember the bright red meat exposed, and the darker red of the blood in the wool, and I remember my father's grim face. "Dogs," he said. His deer rifle was leaning against the corncrib. First he shot the dogs; then he shot the sheep, one by one. A few were already dead, but some of the most grievously wounded were still alive and had been trying to escape the dogs by pulling themselves along on their front legs. One sheep was dead without a mark on her. "Shock, I think," said Dad. He was hanging them to be skinned, butchering them being the only salvageable option.

Nothing was so despised in the country as a dog that killed livestock. A coyote might kill your sheep one or two at a time, but when dogs get started, they don't stop. For Dad this was more than cruelty, it was destruction of property, putting his livelihood at risk. The dogs belonged to the neighbors just up the road. They were newcomers to the neighborhood. Dad went up there to tell them what had happened, and what he had done. He was straightforward but gentle about it. When we first moved to the farm we had a dog named Sam, and Sam had run over to the Andy Dunn place and killed Andy's sheep. Dad has been on the other end of the conversation.

When I recall the look on Dad's face that day, I realize he was facing a serious economic blow. That is hardly the case with our chickens, but man. We *liked* those silly birds. And Little Miss Shake-N-Bake . . . Amy was sad but composed. The killing happened at dusk, so in the morning I took her out and we tried to reconstruct the scene. "He killed my two favorite chickens," said Amy, picking through feathers. I wasn't sure which of the other chickens she meant, but I knew better than to interrupt. "I miss Little Miss Shake-N-Bake the most." She ran in the house, but then returned. "I put two of her feathers in my memory box!"

Jane continues her attempts to convey herself, knitting her brow and squealing meaningfully when we get face-to-face. We are still on a stretch of enforced insomnia as she continues teething. One night I find myself driving to Eau Claire in the middle of the night to buy a tube of Anbesol. By the time I'm back she has fallen asleep. As with any baby problem, we're getting lots of free advice. Some of it we try—for instance, letting her chew on frozen rags. Some we don't try, like the pioneer method of rubbing brandy on the baby's gums—although I'm currently re-thinking that one: After a speaking engagement during which I mentioned the teething and the fact that my wife was at home holding down the fort with a bawling baby, a man approached and introduced himself as a pediatrician. "Here's what you do:

soak a rag in brandy and rub it on the baby's gums . . ." and I thought, Yeah, yeah, but then he said, " . . . and then give the rest of the bottle to Mom!"

Less than a week after the dog attack, we've lost another chicken. One of the Barred Rocks. I was working in the office and saw the birds down around the pigpen. I happened to look up just as the Barred Rock went into the brush behind the trash-burning barrel, and she simply never returned. When the other hens wandered back up to the yard without her, I went to check for feathers but found nothing. A fox? A fisher? A wrong turn? I guess I'll never know, but we have established a 50 percent loss rate. I really need to get that coop done. Mills was working on it the other day without my help. He tried to build a wall but shot himself through the finger with the nail gun. He took a picture of the punctured digit and the puddle of blood and e-mailed it to me. I felt bad for a split second, then mailed him back to check how the rest of the coop was coming along.

Our friend Karen has come over to make sauerkraut. She and Anneliese and Amy are on the deck, working in the sun. Jane is in her baby bouncer. The poor kid, we go about fifty-fifty with disposable diapers to cloth, and today she's wearing a cloth pair that makes her butt look like a cabbage. It doesn't help that she's wearing them beneath a pair of brightly colored stretch pants. I call this her going-to-bingo look, although perhaps I should not. Lately she has developed a drooly gape-mouthed grin immediately recognizable in my baby pictures from the same stage. But her blue eyes, pale as winter sky—those are all Mom.

Anneliese is using a slaw board that was handed down to my

mom from her uncle's mother and has been in our family for
over a hundred years. The board is pretty much just that—a long
board with wooden rails on either side and three deadly blades
mounted at an angle between the rails. You slide the cabbage
head up and down the board, and the blades slice it into strips.
After a century of use the wood is smooth and dark. It got dry
last winter and a corner of the wood cracked. When I came in
from writing late last night, I found it on the table with a note
from Anneliese asking if I could fix the crack. I spread wood glue
over both surfaces, and then clamped the halves together using
a conglomeration of miniature bungee cords and plastic clips.
It looked hack, but in the morning the board held solid and the
crack was nearly invisible. When Anneliese thanked me for fixing
it, her smile was a fine reward, and for the umpteenth time this
year I took note of the fact that I need to review my set list.

Almost immediately Karen cuts off the tip of her finger. I have
just received a new jump kit from the local fire department, so
it's a great opportunity to familiarize myself with the contents of
the bag and review basic bandaging technique. We get the bleed-
ing stopped and I do a serviceable job of dressing the wound.
I do not use clamps or bungee cords. Karen is determined to
continue, and as I pass back and forth through the house for the
remainder of the day the pile of cabbage heads in a cardboard
box transforms into a pile of pale green silage in a crock, and by
the end of the day the kitchen counter is lined with glass jars set
to percolate and produce the perfect side dish to those pigs of
ours.

I am not a deadbeat husband—lately I work probably too
much. But among other things this year is highlighting the dif-

ference between *earning* and *providing*. I should be helping with
that sauerkraut.

The next day Anneliese and Karen can sweet corn and tomatoes.

When Mills and I began working on the coop, the corn was short.
Now it is turning in the field, and my chickens are still home-
less. Oh, but take heart, fowl, because today on a sunny morning
Mills and I met at his place, deconstructed the coop wall by wall,
loaded it piece by piece on an equipment trailer, and hauled it
home to Fall Creek. We are assembling it now as we giggle in the
sun. Before we flop the floor over on its skids, we insulate it from
below with strips of Styrofoam salvaged during yet another one of
Mills's dump runs. I like to think that come January my chickens
will have warm feet. Then we begin remounting the walls. We
get the first one tacked up fine, but then there is a breakdown in
communication ("Slide 'er a tad to the le— RIGHT!! RIGHT!!")
and the eight-foot-tall front wall does a full-on topple, missing
me by the skin of my bald head. It's made mostly of oak, and hits
the driveway with a tremendous thump, blowing dust across my
toes. "LOOOOORD MISTER FORD!" hollers Mills, his hammer dangling from one hand and his eyes and mouth three perfect circles. There is the iconic silent film in which Buster Keaton
stands fast as the front wall of a house collapses over him, and he
survives only because he manages to stand right in line with the
window. Same deal here, except I didn't hang around to thread
the window.

Apart from a few busted boards, the damage is minimal, and we assemble the rest of the walls without incident. Mills roughs out roof boards while I install the windows and tune the doorjamb. The windows were salvaged from my beloved New Auburn house, and it warms my heart to see them put to use. And they bring the structure alive—when I stepped back for the standard moment of appreciation, there was something about the light on the panes that took it from a bunch of boards to a *coop*. When we knock off there's still much to do—tarpaper the roof, install the insulation, mount the roof vent, put facing on the interior walls—but before we quit we cut scrap plywood and nail it over the roof.

Meaning, tonight our chickens snooze in a coop. Sure, it's still sitting in the driveway blocking the garage door, and it's not really finished, but as dusk falls I lure the layers near by shaking a pail of feed, then I sprinkle a trail of it up the cleated gangplank of the door, and sure enough—after much clucking and nervous head-dipping—the Barred Rock pecks her way up the plank and into the coop, where she stands blinking at the new digs. I have to cheat a little with a couple of the other birds, give them a boost, but in short order they're all in place. I rig a divider between the two little doors and then haul the meat chickens over two by two, stuffing them in the second door. When everyone is in place I distribute feed and water, and then before I go into the house, where the kitchen light is now a yellow square in the dark I stand awhile and just listen to the sound of them shuffling and settling.

The mornings are cool now, and knots of color are appearing against the green slope of the valley. I called the farm today and no one answered, and then I realized Mom and Dad were at "convention," an annual assembly of the Friends at a farm an hour due west of here. For four days they will gather on gray wooden benches inside a large white barn to pray, sing hymns, and give testimony. Even this far removed I can feel the peace of it, the cars motoring in slowly while the mist is still clearing the hills, everyone parking neatly in the mown hayfield and making their way to the barn, Bible cases in hand. There will be some lingering and visiting in the yard up until ten minutes prior to service, at which point all but a few stragglers or parents with crying babies will be in the barn and on a bench, sitting in quiet meditation. *Prepare your hearts unto the Lord*, the Bible says, and so it is.

We always opened with a hymn, and it was a great joy to sing at convention, to hear all those voices raised. Even given our constrained ways and hymns titled "We Thank Thee, Lord for Weary Days," the sound of a few hundred open throats does grow you some wings. I remember most of all the older women, the ones with mysterious steamship bosoms and black stockings, and how purely their voices soared. I can't read music per se, but I retained enough from Mrs. North's piano lessons to tell when we were going up and when we were going down. John and I learned to harmonize simply by sitting beside each other and trying to hit notes that seemed to blend. Sometimes they were by the book, sometimes not. But by the time our voices changed brother John and I could manage a serviceable descant as the ladies headed for the rafters.

Anyone who had professed was free to participate in prayer and testimony during the first portion of the service. The bulk of the meeting consisted of sermons from the workers, who rotated through in fifteen- or thirty-minute increments. There was one two-hour service in the morning, another in the afternoon, and a shorter, more gospel-oriented service in the evenings. Mom was realistic about having six or eight kids ride a wooden bench for four days straight; each year we each got a brand-new miniature tablet notebook with a fresh pencil, and during the afternoon meeting she would dole out a few pieces of hard candy to each of us and then—late in the afternoon, and we always anticipated it—a single piece of Trident original flavor sugarless gum. I still buy it just for the memory. When Mom had to leave with one of the babies or to give a tube feeding, Dad's meaningfully raised brow was usually enough to quell any percolating misbehavior. I recall entertaining myself by drawing cartoon heads and watching doomed flies land on the flystrips dangling from the rafters, where they'd buzz in futility until their wings became trapped against the oily ribbon. Some of the boys used to catch flies. Then they'd pluck a long hair from their poor sister's head and tie it around the fly so it could get airborne but not get away. You'd see these flies circling in tethered orbit and some kid evilly grinning. We never got that one past Dad's raised eyebrows.

Between services we ate in a giant tent. It was dark green and probably army surplus. We stood in line, and when the dinner bell rang and the flap was opened we filed slowly inside to the homey aroma of beef stew and piping hot dumplings. We sat at long tables and sang grace, and then the food was served—great marbled plastic bowls of boiled potatoes, plates of sliced peppers and tomatoes, trays of bread and cookies. The whole operation from cook to bottle washer was run by volunteering Friends—children pitched in too, often carrying the pitchers of coffee, tea, and water from table to table, and gathering the dirty dishes as people finished. On cold days the tent was the best place to be, full and warm with the heat generated by all the cooking and the clouds of steam rolling around Mr. Ramsdell in his rubber gloves and apron beside the homemade scalder. The forks made a silver clatter when he dumped them from the basket, the sound of their steely tumble ringing above the muffling green of the trodden grass.

When I grew older, the time between meetings became charged with the nervous hope of love. We were sometimes admonished by the workers to remember that convention was about worship, not dating. But when you belong to a group as rare as this in which marrying outside the faith is fundamentally forbidden and you suddenly find yourself with free time and girls who believe, you make romantic hay. Or try. I rarely got past furtive glances. The standard procedure was to wangle your way into conversation with a likely candidate and then invite her on a walk. The convention grounds were perfectly suited for this, with trails that wound all through the hills and fields, and on a sunny day between meetings they were filled with teenage boys and girls walking in couples and clusters. I envied the boys out there strolling, because I was too shy to pull off anything that straight-forward. In line with church precepts, the girls who walked the paths wore long dresses—mid-calf at the least—that ranged in style from evening wear to *Little House on the Prairie*. Their long, thick hair was wound and folded carefully into buns and swoops held in place with invisible pins and beautiful clasps, and the general absence of makeup lent their faces a frank clarity. To this day the look draws my eye in a way no swimsuit model can manage. I became hooked on the idea of purity, and that hair tumbling down. My poor wife has learned that if she dons an old jeans skirt and twists her hair up to grub around in the garden, I tend to lurk around the kohlrabi and attempt to make small talk.

Of all the children in our family, none have continued in the Truth. Looking out across the sunny country now, over the coloring hills and to the west, I think of Mom and Dad gathered right at this minute, and I wonder if this is heavy in their hearts. Once when it was early in my "losing out," I came to convention with long spiked hair and dressed like a cross between a U2 roadie and Don Johnson's personal shopper. I was sitting by Mom in the dining tent when she quietly wondered what the Friends must think. "I don't *care* what these people think!" I snapped, and she turned her head quickly but I had seen the immediate flash of tears and I was sick with my cruelty. I am still ashamed. But I am better with it now, because although I don't believe, I have never lost the memory of how comforting it was to gather for four days in quiet circumstance with fellow believers. I am happy that they are there, and I hope it is peaceful. Soon enough they will have to come back out among us. Because of the cows, we always had to leave before the evening meal and service. How jarring it was to depart the quiet farm with its fellowship and murmur and shortly be passing by taverns and gas stations and short-haired women in pants.

Today my friend Buffalo came by to inspect the roof of our old granary to see if it would support a rack of solar panels. Actually, Buffalo has informed me they are *photovoltaic* panels, and that if we get some we will be part of the "PV community." Buffalo installs alternative energy systems for a living, and he and his wife Lori were the first set of friends Anneliese and I met and became close with as a couple. Although I am happy to say we each get on well with the friends the other brought to the marriage, it is also nice to have "shared" friends, and it doesn't hurt that they have two daughters roughly Amy's age. Anneliese invited them over today under the pretense of dinner, but in addition to spec'ing the granary, I've bamboozled Buffalo into helping me finish off the coop. While I cut and staple insulation between the studs, Buffalo tarpapers the roof and cuts a hole for the roof vent. After he helps me install a row of plywood facing around the base to keep the chickens from eating the insulation, I bring the tractor around and we make the big move.

The tractor moves across the yard with the coop in tow. In a rare moment of foresight, we removed the windows so they wouldn't bust in transit, and Buffalo is riding crouched in the window waving at the kids like an underweight troll, his head of curls and big black beard flopping in the wind. For my part, I keep 'er steady with the tractor, one arm raised and pointing to the distance as if I am Hannibal headed for the Alps. The three little girls dance and wave from the deck.

We pull the coop into a patch of weeds beside a chokecherry tree, the windows facing south to catch the winter sun and allow the hens a view of the valley as they squeeze out their eggs.

When we head to the house for supper I notice the coop is sitting at a pretty good angle, but it looks solid there on the horizon, just like I imagined all those months ago when I was poring over schematics drafted in 1933.

In the morning I rig a fence for the meat chickens. One of them developed splay-leg a week ago. I tried taping its legs together like it said to do in the chicken book, but he didn't get any better. He couldn't walk, so I put him within reach of the food and water, but the other chickens stampeded over him. "That's because chickens are small in the head," Amy said. Over the course of several days he declined, and today I find him dead, which makes me think I should have knocked him in the noggin early out of mercy.

Three days after the move, I have to leave to participate in a literary festival, but it is only a few hours from here and the hosts have graciously offered a place for the whole family to stay, so we're turning it into a mini-trip. We have arranged for Anneliese's sister Kira to watch the livestock. I still haven't quite got all the layers trained to roost in the coop instead of the pump house, so to save Kira the trouble of rounding them up at night I decide to rig a makeshift fence. While trying to finish the fence in a rush the same morning we are leaving, I manage to knock the roll of steel chicken wire over just as one of the layers is making an inquisitive pass. It's one of those slow-motion moments where I can see the heavy roll falling and the chicken boop-a-dooping along right into its path, and sure enough even as I lunge for the roll it falls *whump* right on top of the chicken. She looses a horrid squawk and runs off when I lift the roll, but she is limping badly. Before we depart, I type up a letter of instruction for Kira and leave it on the kitchen table. It gives a fair summary of our progress here in the Year of the Coop:

Hi Kira:

First and foremost I shall apologize for (A) the cobbled-up state of my chicken operation, and (B) the length of this set of directions, which far exceeds the complexity of the tasks at hand and will take longer to read than the time required to actually care for the animals.

PIGS

They oughta be fine, really. I have put feed in their feeder and they have water. If you have time on Saturday, lug a pail of goat milk (in the fridge behind first garage door —keypad is broken, use opener on windowsill over kitchen sink) down and pour it in their handcrafted blue plastic trough. Then watch in wonder as they snarf it down, sometimes blowing bubbles out their snoots. If you have some spare sweet corn, throw it at them. You are allowed to scratch the pigs if you wish. There is a scratching stick conveniently placed near the fence for just that purpose. Please do not attempt to ride the pigs or take them to town for tattoos and piercings.

CHICKENS

I have done my best to eradicate all the chickens but have attained only about a 50 percent kill rate so you still have to feed the survivors. One black hen might be limping because today I—wait for the irony here—dropped a gigan-

tic roll of chicken wire *on top of her. You really have to aim, and even lead them a little in order to do that, which is not easy with a heavy roll of* chicken wire.

The white chickens are basically meat cell replicators on legs. Pigs with feathers. Dumber than a box of, of, of, well, feathers. In the morning, open their little flap door and set them loose to run wild and free (within the confines of their fence). To access the compound, undo the netting where it is held in place on the coop with a drywall screw. You can load their feeder right up, they'll eat at it all day. Replenish their water. The hose is strung right over there. Don't forget to turn the water off when you're done because it leaks and also Anneliese has a little lecture she gives to people who leave the water on, if you'd like I can recite it by heart. I usually put their water and feeder out during the day but you can probably leave it in, that way if it rains the feed won't get all sogged up. The feed bag is right inside the coop. There should be a little cup inside the bag but I might have forgotten to put it there. Fill the feeder up again at night, please.

The layers are mad at me because I penned them in today. They like to run free, but by penning them up I save you the trouble of running after them at dusk. It's a nice little evening frolic, and I rather enjoy it, but then that's me. So just turn 'em out in the morning, reload their feed and water (you probably gotta reach in under the netting divider to get to the feeder and waterer), and shuck 3–4 cobs of sweet corn (under the white plastic bucket under the brick) for them. If you think of it and have time, take *the rake and give them some lawn clippings. It keeps them occupied and lends an earthy flavor to the eggs they don't lay. When you do open the gate to open and close their coop door, keep an eye on them: they are alacritous little buggers and will shoot right out on you. The gate as you will see is held in place by tension and two nails. The chickens tend to put themselves in at dusk, and should be waiting for you to close the door.*

That should cover it. Call me at any hour with questions. There are no dumb questions, Kira, only questions that make me chuckle condescendingly.

We are very grateful that you are willing (or guilt-laden enough) to help us out like this.

I usually get paid by the word.

Pig-butchering day dawns crisp and sunny, cold enough that the chickens are fluffed out on the roost. The coolness makes it a good day for handling meat. I'm happy for that, but as I work around the garage—bagging the garbage, sorting the recycling, wrangling plastic pails for the morning's bloody work—I keep catching the pigs in my peripheral vision, and I'm surprised at the leaden patch of dread in my gut. This day was booked the second I wrote the check for those pigs, and when I brought them home I was bringing them home to be butchered. It has ever been the intent, and that won't change, but until the sun rose this morning the concept existed in the abstract. The pigs are not pets. They have taken just enough nips at me that I know they would afford me no courtesy given the chance, and I watched them crunch up that rabbit, but still: I liked having them around. They were genial in their grunty, hoggy ways. I'm musing along like this when I hear the truck coming, and once Muzzy hits the yard, all introspection ceases. He roars right past me, straight down to where the hog pen is, and by the time I'm down there he's already backed around and is stripping out cable from his hoist. The pistol is holstered on his belt. "Let me know what I can do to help," I say. "I could use a bucket of water," he says. I jog back up to the house, the young kid eager to please. I'm waiting for the bucket to fill when I hear the *pop!* of the first shot.

I've told Amy it can be up to her if she wants to help with the butchering, but I didn't think she should be out there for the kill-ing. I'm not sure why—she loves to go hunting with me, and last year she was at my side in the tree stand when I shot a deer. She tromped through the swamp and then sat there patiently in the cold for two hours without sound or complaint, and when I shot the deer, I put her on the blood trail and she followed it right up to the carcass, at which point she exclaimed, "Oh! A nice plump doe!" But these pigs—I don't know.

When I return with the water, both pigs are down and bleed-ing out. The thing with pigs is you have to lance the carotid as soon as they hit the ground, or they flop around horribly. They're back in the paddock a ways, so I get the tractor and a chain to move them out to the truck to save Muzzy the trouble of thread-ing the cable down through the gate and fence posts. Muzzy has hooked the hocks on a length of steel with a large eyelet at the center. While I feather the hydraulics, he hooks the chain to the eyelet, then takes a couple of wraps around the loader, and I back out of the paddock, the pig swinging head down until I lower it to the grass beside the truck, where Muzzy dives in with his knife, severing the front legs.

"Where's your daughter?" he asks.

"In the house," I say. "I figured that was best."

He pauses with the knife in midair, looks straight at me. "Well, I'm going to tell you I think that's a mistake." He lets the statement hang there a bit. "She should see this. Kids can learn a lot from this. Pig's very similar to a human. Sometimes I take the guts and eyeballs into the science class at the school, so they can study 'em."

He goes back to cutting, and I go to the house. Amy and Anneliese are working on a homeschool lesson. It turns out Amy has watched the slaughtering from the upstairs window anyway. They are both a little unnerved. I ask Amy if she wants to come out. She looks at her mother. Then she looks at me, a little hesitant. "It's up to you," I say, "but the man thinks you might learn some things."

"OK," says Amy, and so we all trek out. Jane is sleeping, so Anneliese takes the portable monitor.

Almost immediately Amy is engrossed. "What's that?" she says, pointing to the rumpled skin of one pig already flopped in the loader bucket. Muzzy works fast. "That's the skin," I say. "See the bristles?" "Oh," says Amy, then, "Ooo, look at the eyeballs." And then it is full-bore biology lab. Muzzy's knife flashes as he circles the pig, the skin falling in a drape as he works. When he gets ready to split the belly, he stops and gives us a serious look. "Who's the shortest one here?" Amy raises her hand. "OK!" says Muzzy. "You get the job!"

"What job?" says Amy.

"I need someone to crawl in there and push the guts out!"

Amy looks at Muzzy, then looks up at me. "Whaddya think?" I say, breaking into a grin. She looks back at Muzzy; then you can see it dawn on her that he is joking. "Noooo!" she says, giggling. Muzzy laughs happily and starts peeling the guts out. But he keeps stopping, using his knife as a pointer, urging Amy to get closer, to have a good look. "See that? That's the spleen!" He cuts it free and splits it, points out the vascularity, tells her how it can be injured in a car crash. He gestures toward the underside of the liver. "That's the gallbladder!" Amy is fascinated. When he shows her the heart, he explains how it works and how a pig heart is like a human heart. "Sometimes they use parts of pig hearts in people," he adds. Amy is soaking it in. He lays the lungs on the ground and dissects them, showing her how the air goes in and out. Then he asks, "Are you going to smoke cigarettes?" Amy shakes her head solemnly. "If you smoke, your lungs will have all kinds of black spots inside them," Muzzy says. Then he slings the lungs in the bucket. They land with a slickery flop.

When it is time to halve the pig, he produces a monstrous steel hacksaw and plugs it into an inverter outlet on the truck. When the pig is split, he rotates one half out to show Amy how the brain lies tight in its case. She squats down and has a good look. "I can see his teeth!" she exclaims.

I have backed the pickup down to the butcher site and lined it with plastic. Muzzy swings the winch boom over the bed of my truck and slowly lowers the pig as I guide the halves into the bed. Then he starts in on the second pig, Amy at his elbow from start to finish. Muzzy continues in the professorial vein, but we also get him going on stories. He has been working the entire time with his fingerless hand stuffed in an athletic sock, the thumb protruding through a hole in the fabric. He keeps a small meat hook pinched between his thumb and the palm of the hand, snagging the meat and skin of the pig as need be to set up the cut. What the heck, I think, and just plunge in: "So what's the story with the missing fingers?" In more delicate company I might have anticipated gasps and umbrage, but Muzzy launches off as if he thought I'd never ask.

"Corn picker!" he exclaims, almost triumphantly. "Took these three fingers right off, and degloved this one." I've seen a few degloving injuries in my day. I know he would have seen the bone sticking out naked as a Halloween skeleton, the skin stripped away. "So they amputated that."

"What about your thumb?" I asked. "Is it a toe?" The thumb looked a little flat, and I know they do that with toes.

"NO!" says Muzzy, emphatically. "They tried to do that. But I told 'em, I *need* that toe. I was a truck driver at the time. You use your toes all the time—shifting, pressing the gas . . . Nope, I wouldn't let 'em take the toe.

"I was in the hospital for eight days. Four days after I got home I was back to butchering." The palm of the hand looks padded, like a mitt. "They took meat from my forearm to build it up," says Muzzy. "Then they covered it with skin they took from my leg. My leg hurt worse than anything else."

His thumb looks cold in the sharp air. The white sock is wet and reddish with blood. "No, the sock keeps it warm," he says. "Cold's not the problem. It's got good circulation. Cold don't bother it." He looks at me with a reckless grin. "But you stick it in a bucket of hot water, and I'll go to the moon!"

When both carcasses are in the truck and Muzzy is gone, I pull the plastic sheeting around the pigs and drape a logging chain back and forth over the plastic to keep it from blowing loose. Then Amy and I drive the pigs north to Bloomer, where we will turn them over to my friend Bob the One-Eyed Beagle. "Tell him to save the fat," says Anneliese as we are leaving. "I want to render the lard."

I delight as usual in having Amy as my copilot. Bombing down a country road in a pickup truck with my daughter has become one of the signal joys of fatherhood. Throw a couple of dead pigs in the back and you've got yourself a Hallmark card on wheels.

Trucking the carcasses up north is clearly a violation of local food principles, but loyalty trumps all, and the Beagle and I served on the same fire department together for over a decade. Furthermore, he is a good citizen and a fine butcher. And finally, only a short-sighted churl would pass on the opportunity to haul his homegrown pigs from a one-handed butcher with two eyes to a one-eyed butcher with two hands.

After we unload the pigs and see them swaying on their hooks, Beagle gives Amy a tour. She gets to see the knives and saws and the cold steel tables, the massive half-cows hanging. She takes it all in, and turns to look at me with her eyebrows raised when the Beagle gives her three guesses to identify the one carcass different than all the others, before revealing that it is a skinned bear. While the Beagle demonstrates the vacuum sealer, I think how grateful I am for those friends of mine who can do more than just sit and type; these friends who have fundamental skills and trades reflected in the condition of their hands. It is good, I think, that in the hubris of the digital age this little girl be given a look at the more gristly bits of existence.

We're just nine miles from New Auburn, so I drive up to the farm to visit Mom and Dad. Mom says there is a chance of an early first frost tonight, and their neighbors Roger and Debbie need help getting in the last produce. Mom joins Amy and me in the truck, and we drive the dirt road to where Roger and Debbie have their fields in the pines. We load their Gator with pumpkins and squash and watermelon and gourds, and when it is full, Roger makes the run back to the shed.

Before we leave Roger and Debbie encourage us to take whatever we like. We load up a little bit of everything and then head over to help my brother John and his wife Barbara pick all of their pumpkins and get them under cover. John and Barbara supply our family's jack-o'-lantern carving party every year, and some of their pumpkins are such monsters that John uses the skid steer to shift them. We've picked up Amy's cousin Sienna on the way over, and the two of them are yakking and scampering through the vines, picking pumpkins small enough for them to lift and carry over to the pile. When all the pumpkins are picked and tarped, we head back to Mom and Dad's again. We need wood shavings for the chicken coop, so Amy and I go to the lumber shed and fill several bags from the fragrant mound of piney curls beside the planer.

Back in the house, Jed has come for Sienna. We wind up in the living room where we wrassled when we were kids, only now we sit in chairs and just talk, and talk for a long time. A little bit about how he's getting along, of course, but also a lot about nothing in particular. When I look back at this day—from rising early to prepare for the pig slaughter right up to this easy moment—I wonder at how much can be had when there is no clock in sight, no destination pending. On the drive home it is cold enough that I turn the truck heater on, and by the time we pass Bloomer, Amy is asleep.

CHAPTER 10

Sunday mornings when I was a boy I worshipped the Lord in a white clapboard house. I sat in a straight-backed chair against a hard plaster wall. In the summer the plaster was cool, and in the winter it was cold. The windows were narrow and tall, and the glass was ripply—a distracted little lad could rock in his seat and roll a shimmy through the trees. Sometimes when the snow was heavy on the ground and the living room was radiator warm, the boy got drowsy in his sweater and corduroys. When his head lolled back, it rang soundly on the plaster so that the windowpanes—resting loose in their fractured putty—buzzed like snares, the racket signaling that someone was snoozing along the path of righteousness.

The lady who owned the white clapboard house emigrated to this county in a Conestoga. Having read many cowboy books, I knew Conestogas were for pioneers. I would study the frail woman sunk in the worn chair with her Bible across her knees and thrill to think that once she was a young girl peeking from beneath the flapping canvas of a prairie schooner. Imagine: to worship the Lord beside a pioneer! I can no longer conjure the lady's face. I remember she had yellowish white hair, which she wore pinned atop her head, as did all of the women in our church. I remember the voice of her son the church elder guiding us quietly from hymn to prayer to homily, the King James cadence of *thee* and *thou*, of *shalt* and *wilt*, and in the winter the big Jungers furnace in the corner with its blue flame wavering. For a hallowed hour this house was holy, and we were the chosen ones, cradled separate from the world.

I loved our church of no churches. I loved the little white clapboard house.

I did not doze off because I was bored.

I dozed off because I was cozy.

Within the house there were the chairs against the wall, and a row of folding chairs before those. We had no assigned seats, but tended to gravitate to the same place every Sunday. There was Florence beside the Jungers, and her sister Vivian beside her, and their stepmother Myrtle—all three of them sturdy women who sometimes fished lace hankies from deep within their edificial busts, an utterly asexual move that nonetheless widened a young boy's eyes. Along the wall, sunk so deep in an ancient wine-red velour couch that their kneecaps came to chin level, were the three Jacobson boys, teenaged grandsons of the woman who came here in the Conestoga. Mrs. Doury's granddaughters and daughter-in-law sat in adjacent chairs, and depending on the Sunday, a handful of other worshippers might round out the congregation. Mrs. Doury's son-in-law John served as elder, meaning he directed the service. The bread and wine (grape juice and a slice of Wonder Bread) sat beneath a white handkerchief on an end table at his side.

We sat quiet until exactly 10:00 a.m., and then John spoke.

"Would someone like to choose a hymn?"

I was always hoping for Hymn Number 1, "Tell Me the Story of Jesus," because it was my favorite and I knew most of the words without looking, but it was usually reserved for gospel meeting. So someone suggested a number, and we paged to it in *Hymns Old & New*, and then one of the women—Florence, usually—would lead the singing, hitting that first note so the rest of us could follow in behind. Once she chose the range, you were stuck with it. Sometimes you'd have to drop an octave to hit the high notes and jump an octave to hit the low ones. When the first hymn was complete sometimes we sang another one. Then John said, "Let us bow our heads in prayer."

The prayers rose around the room in no particular order, with the exception that John the elder always went last. The prayers were usually brief and simply worded: *Lord, we pray that thou wouldst grant us stillness in our hearts; That thou wouldst improve our spirits; That we might find ourselves worthy of thy mercy.* Some prayed in a rush, some prayed briefly; some prayed a different prayer every Sunday, some prayed the same thing every week. By and large the prayers were poetic in simplicity in rhythm, and everything remained resolutely in the spiritual realm—overly specific requests were seen as unseemly. (Thus I was quite unprepared later in life when I overheard a prayer session among a group of young evangelicals at a local coffee shop during which a young woman quite fervently prayed, "Lord, you have got to get me out of this lease!") Throughout the time of prayer, we children were expected to keep our heads down and eyes closed. I do remember sneaking peeks, although not often, because somehow even with his own eyes closed, Dad would catch me and I would get the eyebrows.

When the prayers concluded, we sang another hymn, and then it was time for testimonies, when those who chose to participate shared a Bible verse or several that they had been meditating on during the week and then offered a homespun homily. The first time I gave testimony it took a while for me to get my gumption up, and I quaked as I said, "My thoughts this week have been on Matthew, chapter 19," and then I read aloud verses 16 through 22, in which a young man asks Jesus what good things he must do in order that he may have eternal life. Follow the commandments, replies Jesus. I have done that, says the young man. Then sell all your possessions and give them to the poor, says Jesus, and the man leaves in a funk, as he has a great number of possessions. "I hope that I will always live so that I am storing up riches not for this world, but for eternity," I said, and then the next person began to speak and I felt great relief. As with prayer, the testimonies moved around the room in no particular order, and then when all had spoken, John the elder gave his testimony. When John concluded, he set his Bible aside and said, "Would someone give thanks for the bread and wine?"

Once you professed in gospel meeting, you were allowed to pray and give testimony Sunday morning, but in order to take the sacraments you had to have been baptized. We were of the Anabaptist persuasion, eschewing infant baptism, trusting instead that when the Lord so moved us as believers we would seek out the workers and request to be included at the next baptism, a full-immersion ceremony usually held in a river or farm pond. I never got baptized and therefore never "partook of the emblems," as we used to say. I do remember helping pass them around the room, and how heavy the glass felt, and how I focused intently on handling it so as not to spill it, and how it seemed imbued with a heaviness far exceeding a glass of juice. As the emblems circled, each baptized person took a pinch of the bread and a sip of the wine that wasn't wine. Following the bread and wine, we sang a final hymn, and church was over. It rarely went over an hour. We rose and shook hands all around. You made sure you got everyone, young and old. It was an informally required formality. The grown-ups visited, and then we sorted our hats and coats from the pile in the kitchen and stepped back into the outside world for six more days.

When I went to help my sister and brother-in-law butcher their chickens, the thinking was that I would build up some sweat equity credit and they would help butcher ours, but now they won't have to. Our neighbor Terry—with whose family we split the original meat chicken order—has arranged to have his chickens butchered by a local Amish family. They can process up to fifty chickens per day, and since Terry has to haul his bunch over there anyway he asks if we'd like to buy a ticket to ride for our seventeen (originally twenty: one DOA, one terminal splay-leg, and one crunched by the chicken tractor). At first I hesitate, strictly out of hammerheaded pride (turning my chickens over to be butchered by someone else impinges on my delusions of self-sufficiency), but then I visualize the leaf pile of bills, Post-its, and rough drafts covering my desk, and it hits me that sometimes delegation is the better part of valor. Later I bolster this line of thinking with the justification that we are supporting the local economy, although to what extent I am not sure since we will be charged only two bucks a chicken. All arrangements have to be made by mail, so once we commit, Terry sends a letter to a man named Levi confirming the date and time and number of birds, and we mark the calendar.

The birds are ready. They have grown at a steroidal rate on their hog feed and now clomp around the confines of their pen like clucking sumos. As a man I think the bigger the better, but Anneliese finds their growth rate unnatural and would like to try raising some leaner heirloom varieties next year. We have managed to veganize them slightly by shifting their fence so they can peck at greenery, but they still lack the verve for foraging that we see from the layers, who frequently hit the yard in a flying wedge, driving autumn's last grasshoppers before them like desperate fleeing popcorn.

I no longer believe all I believed when I sat in my chair in the white clapboard house, but I am not prepared to scoff. There is enough derision in the world. That is not to say I am above knee-jerk crankiness. When a stranger on a bus asked if I was a Christian, I shot back perhaps a little too sharply, asking if he would treat me differently depending on my answer, and I could see immediately he hadn't meant it that way. Because I grew up worshipping in a manner that could be described as unplugged and acoustic, I sometimes wax a tad irascible about churches that serve lattes and happy music. Church should not be easy, I said once while giving a talk, church should be *hard*. After which a woman mailed me an envelope containing full-color photographs of Asian children whose tongues had been ripped out after they professed Christianity. You see, the lady wrote, church *is* hard. Clearly we were talking past each other by the width of several zip codes. I get most crotchety when someone proselytizes me with an aura of patient indulgence, as if I am a fuzzy-headed wandering lamb who scampered off to the devil's clover patch one day and never looked back. Just because you drop the dogma doesn't mean you don't dread the price of transgression. Mine is a chastened apostasy—I don't claim to have the answers, and although I stand outside the church of my parents, I still peek through the windows for guidance.

I wasn't surprised when Amy asked me about God. All children get around to it. It was actually a couple of years ago, and her voice came out of the darkness behind me in the van. I was stumbling through a wish-wash mumble when she granted me reprieve by interrupting with another question: "Why did the men kill Jesus?" That one is easier, because I may be a silly wandering lamb, but the story of Jesus, that one is written on my heart, every word. So I talked about Jesus. About how he lived, what he taught, and how he died. And then she asked if she could have a horse, and I was given time to regroup. But I realized the future had arrived, and there would be no opt-out. You cannot toss your seven-year-old a copy of *Being and Nothingness*. As eternally dangerous as it might be to end up a bumbling agnostic, it may be even more dangerous to *begin* there. The greatest gift my parents ever gave me was a firm foundation.

Lately we have begun going to church.

We have not made any final decision. For a while we attended services at the local Unitarian Universalist church. Then we went to several Quaker services. Currently we are attending Mennonite services held in a Jewish temple. We have felt warmly welcomed in all three settings, but neither Anneliese nor I have been able to settle. I bring with me my prejudice against anything more organized than a camp meeting, whereas Anneliese—with her background in Lutheranism, Catholicism, and shamanism—finds herself longing for more ceremony. None of this is made easier by the fact that we have friends and acquaintances in all three congregations, and there is the usual polite desire to keep everyone happy.

A couple of times already when I have been behind on some deadline or another, I have stayed home while Anneliese and the girls go to church, and this shames me. When I was a child, Sunday was a day of rest. The Ten Commandments, you know. No matter how far behind he might have been, no matter how much hay was down with rain threatening, Dad saw to only those chores required for the comfort of the animals, went to church, and took the rest of the day off. One year when incessant rain was blackening the mown hay in the fields a rare sunny Sunday dawned, and Elder John took counsel after the morning meeting to see if he could justify baling hay that afternoon. With forage desperately short and more rain coming, it would have been wasteful to let it lie, and so they baled, but I remember thinking it was a momentous decision. Issues of God and faith aside, I am thinking my little girls should come to see Sunday as a day apart. As a day to set all worldly business aside and abide in ceremony. I will never cut it as a Quaker—I cannot find it in me to renounce all violence, not with two daughters under my protection—but I do love their silent hour, which in my case invariably evolved into a self-scouring meditation on the idea that the busy life is not the full life.

For better or worse, I have to play it straight with the kids. When Amy was four she woke up three nights in a row screaming that monkeys were flying in her window. That third evening she was being watched by a babysitter, and the following morning Amy said the babysitter told her Jesus would make the monkeys go away. That night the monkeys were back. How do you finesse that one?

Despite the depth of my parents' faith, they never oversold the church. Two years ago I asked Dad about the origins of the Truth. "The workers will tell you it comes directly from God," he said. "Actually it came from Scotland. Sometime around 1900." After a lifetime of watching him walk so faithfully, the honesty of his answer floored me. Later Mom confided that after years of being assured by the workers that the Truth could be traced straight to the twelve apostles, the discovery that the sect was actually the offshoot of a group formed in 1897 by an itinerant Scottish evangelist named William Irvine did indeed leave them feeling deeply betrayed, but the one issue that nearly drove them out was their refusal to condemn people of other faiths. "We did not, do not, and will not," says Dad, before going on to list friends, neighbors and acquaintances whose spirit he admires. When Mom and Dad were confronted by a worker over their dissent, Dad invited the man to throw them out. It didn't happen.

I have only recently (and mainly because I am now responsible for two children) begun discussing many of these issues with my parents. My hesitancy is rooted mainly in simple respect. Having watched how my parents have lived their lives, I have no appetite for spiritual fencing matches. And although I doubt that I could, I have no interest in derailing gentle people. I do not discount Romans 14:13: "Let us not therefore judge one another any more: but judge this rather, that no man put a stumbling block or an occasion to fall in his brother's way." The lapsed believer does not shed the vestiges of doctrine.

But I'm glad we're talking. During one recent exchange I said Mom and Dad's refusal to condemn "outsiders" (Dad avoids the term, saying it has a ring of arrogance) made them to some extent skeptics within their own church. No, Dad said. Mother and I have *misgivings about the church*. We have no *skepticism* about God and His Son. And it struck me then that if none of us followed our parents in the church, perhaps it is because they refused to follow it blindly themselves. Their actions signaled to us that as important as it was to live in "the Truth," it was more important to live truth*fully*. Before their children above all.

Because of their example, I am slowly turning the corner on why even some skeptics stick with church. "Men are better than their theology," said Emerson, and while I can't see going back, I will be perfectly happy—perhaps even relieved—if my girls become Quakers or Catholics or sister workers—as long as they treat themselves and others with care.

Amy still asks me for stories from my childhood. She's done it often enough now that it sometimes takes me a while to generate one she hasn't heard before. There is the sensation of opening a dented recipe box to riffle through dog-eared index cards. But I dredge one up every time, because I know the inexorable hour approaches when the star power of the yammering bald guy will wane and sputter to nothing. Tonight when she asks, we are tooling down the darkened highway in our dilapidated fambulance, so I tell her about the time our second secondhand Volkswagen bus broke down on a winter night when we were on our way home from gospel meeting, leaving our double-digit family with no ride but the farm pickup. The next time we went to church Mom, Dad, and the toddlers crammed into the truck cab while the rest of us wrapped ourselves in sleeping bags and rode in the back. Dad bolted plywood sheeting over the bed to shelter us from the wind. Unable to sit upright beneath the plywood, we lined the crawl space with old couch cushions and lay on our backs, making a game of trying to judge our progress by tracking the turns, imagining our bodies as needles spinning on a compass.

"Tell me *another* story from your childhood," she says when I finish. So I tell her about the little boy who fell asleep in church on Sunday, and she giggles. Surely she is filing a few index cards of her own. One day she will draw one composed on this night, about her benignly freakish parents and how they dragged her around in a tatterdemalion van that smelled of pig feed and home-brewed goat cheese. And then she begins to sing: *O Lord, prepare me to be a sanctuary, pure and holy, tried and true* . . . It is a song her mother taught her, and her voice hangs in the air with the purity of starlight.

Terry arrives before dawn, and like thieves we load the chickens into his trailer. They are thickly feathered and hefty in my hands and armpit warm where my thumbs stick beneath their wings. Because we make our raid early we don't meet much resistance, and when we pull a tarp over the trailer only a few disgruntled clucks seep through the canvas. Terry tells me the Amish family will be expecting me at five that evening, and then he drives off, the trailer lights stoplight red in the dark yard. I should be butchering those chickens, I think, one last time, then I console myself with the knowledge that Amy, Anneliese, and I have been butchering our own deer on the kitchen table for three years now, and then I review the mental jumble of other things I can do today other than pluck and gut seventeen chickens, and I think, *Well, OK.*

It's cold, gray, and windy when I drive into the Amish family's yard ten hours later. As I turn around and back up to the trailer where it is parked beside the house, several straw-hatted heads pop out of the woodshed. Young boys at work. When I have some trouble adjusting the hitch, one of the boys scurries away to a shed and returns with a wrench and hands it to me silently, but then when I still struggle they jump in to help manfully, the way young boys do when they want to demonstrate their abilities. Finally the cup clunks in place over the ball and I lock the hitch in place, then let myself into what I assume was the garage before these carless folk moved in. Now it is a large room jammed with countertops on sawhorses, galvanized tubs of water, coolers, tubs of chicken feathers, and some fourteen pint-sized chicken pluckers—barefoot children in long dresses and overalls, working beside several adult women and teen-age girls. When I walk in, the smallest children draw toward the women and look furtively from behind their skirts. There is only one man—Levi—and he greets me with a smile. "We are almost finished," he says. Some of the coolers are already packed, so I begin loading them. The women are still bagging chickens—long gone are the stolid white-feathered beasts of the morning, replaced by pale yellow carcasses, headless with naked pointy wings and their drumsticks neatly trussed. In one galvanized tank the carcasses float in the water like giant waxy bobbing apples.

The little boys hustle to help me tote the coolers, clustering busily, heaving and ho'ing in a further attempt to prove their mettle. By the time we get the coolers in the trailer, the women are bagging the last of the chickens. Levi and I review the hand-written bill and tally, and as I write the check the little girls in bonnets peer up silently from behind Mom, and then I am back in the car and away.

Back home after supper and with the baby asleep in bed Anneliese and I get out our vacuum sealer and start sealing pieces and double-bagging whole birds. The carcasses are huge—a couple of them top eight pounds on the kitchen scale. When we feel we have enough whole birds bagged, we clear the kitchen island, round up knives and cutting boards, and start chopping the remaining birds into pieces, removing the drumsticks and wings, fileting the breasts, saving the backs for stock. Amy is already up past her bedtime but she is so eager to help, we tell her she can stay up thirty more minutes. She tucks in happily, sawing away at wings and thighs and helping push the button that runs the vacuum sealer. When the half hour is over she slumps a little but we hold the line, following her upstairs to tuck her in and kiss her and thank her for helping.

Then it is just Anneliese and me at the island, cutting and talking and sealing. I have a chance to look at her in the light and consider us together, and there is much in the year that has gone off the rails or been pushed aside or lost in the hurry, but here we are, putting up stores for the winter. When the last bird is chopped apart and sealed, we carry the cardboard boxes of meat to the chest freezer in the garage. The freezer is already filled with bacon and pork chops and pork roasts and a pair of hams the size of a tortoise. Now as we work shoulder to shoulder finding spaces for all the chicken, it feels good, like we are yoked together not just in workaday dray but in fulfilled purpose. When the last bird is stashed, we step back and look at the freezer, lid up and full to the rim with meat—every bit of it raised within a hundred-yard radius. Standing there beside my wife, both of us in tattered flannel shirts and grubby jeans, tired and our noses wet with cold, I pull her close and for a long moment we just stare at the freezer, and later we both agree it was one of the most oddly happy moments of our marriage since the exchange of vows, because we did this *together*.

Nearly every evening around suppertime I am reminded that John Menard is worth $7.3 billion and I am not. The evidence comes hissing from the clouds in the form of one or the other of Mr. Menard's Cessna Citation Bravo jets returning the managerial troops from their business at the multitudinous home improvement stores, lumberyards, and distribution centers he

owns all across the land. Last I checked, a *used* Cessna Citation
Bravo will run you well north of seven figures. Rather than be
disturbed by the jets (in fact the fleet docks at an airport eleven
miles distant, and although we're frequently in the flight path,
the craft are still at a relatively unobtrusive altitude), I find them
a fine source of existential calibration. I pause in what I am doing, tip my head back, watch them slice the sky like barracudas
on the wing as I ponder current rates of exchange, and then I ask
myself: "So, Mike—how'd *you* do today?"

In taking my measure in this manner I am following in the
footsteps of my father, who has a long tradition of fruitlessly com-
peting against Great Men of Industry. For a while it was J. Paul
Getty, and lately he says he's closely shadowing Warren Buffett,
but back in the 1970s it was shipping magnate Daniel Ludwig.
"Gotta go catch Daniel Ludwig," Dad would say as he pushed
away from the dinner table for yet another round of chores. At
the time, Ludwig was considered the richest man in the world.
Dad pronounced his last name "Lewd-vig," which always tickled
us kids. When one of our milk cows gave birth to a scrawny bull
calf, Dad named him Daniel Ludwig and announced that this
was the calf that would finally let us get ahead. Given the price
of bull calves, this would have been a joke in any case, but in an
ironic twist not only did Daniel Ludwig the calf fail to thrive,
he failed weirdly, remaining skinny and rickety and sprouting
patches of creepily silky hair. By the time Dad shipped him, he
had gained almost no weight but had grown a pair of gnarled
mutant horns.

On better days when we hustled right till dark and got the last
cornfield cultivated or one more load of hay in the mow before

the thunderstorms hit, Dad would dismount the tractor or tamp the final bale down, then stand in his baggy overalls and cracked leather boots, and happily declare, "Now we're catchin' Daniel *Lewd-vig*!" The grin on his face was a wide-open acknowledgment that it wasn't ever going to happen.

During the summer I promised Amy we would pitch a tent and sleep outside one night. Now we're getting frost and I'm about to go on a book tour and play some band dates, and the sleep-out is not going to happen. I love life behind the wheel; the road is filled with friendly faces, and the events support our little family. And while many of my friends, relatives, and neighbors are being deployed into harm's way again and again, I am driving my Chevy to a nice bookstore in Oskaloosa. But what I suspected at the beginning of this year is true: if a man is away from home nigh unto one hundred days in a year, he will wind up doing things in passing. And you can't farm *in passing*. You can't be a good husband *in passing*. You certainly can't be a good dad *in passing*. On my desk is a list Amy scrawled in pencil the day we planned our campout:

> *FOOd. watr.*
> *tea*
> *camPStove*
> *Flansh light*
> *sleeping bag*

Many nights after milking, Dad played softball in the cow pasture with us. We used milk replacer bags for bases and rotated available kids in from the outfield to take their turns at bat. When Dad batted, John and I ran back to stand against the woven wire fence, but it rarely did any good, as Dad would snap the bat around and drive the ball high into the white pines, where it would tumble down in increments, clunking off the big limbs and snapping twigs. We played right through dusk and into the dark, until the ball was just a gray smudge and the dew was fallen. Some nights he took a few of us fishing in the canoe. I can remember him paddling back across Bass Lake in the dark, the smell of the warm water, the sound of the Hula Popper smacking the lily pads. Years later Mom told me that many of those nights he couldn't feel the paddle in his hands, his carpal tunnel was so bad. He would be up half the night in pain, with the next day due to start before the sun. By the time the last cow was milked the following evening, he must have been aching for sleep. And yet he made time for us. "If you tell your child you're going to build a treehouse, build it," says the writer Jim Harrison, "or you'll live forever in modest infamy." Amy's camping list is in clear view on my desk and will remain until she and I spend a night in that tent.

I am on the road, half a state away with my usual trunk full of books. My cell phone rings. It is Amy, her voice brimming with excitement. "Guess what I am holding! Right in my hand!" I play naive. "A toad?" "No! I just got it! It's still warm." I hesitate, generating the next wisecrack, and she can't wait any longer. "An egg!" "No way!" I say. "No! Really! It's still warm! It's brown!" "Well, that's wonderful, Amy." And it is. The egg cupped in my little girl's hand is the tangible result of conversations held clear back before Anneliese and I were married. And the eager pride in Amy's voice reminds me of what Anneliese often stresses—that we are doing these things as a family. Even if I did spend the night in a Super 8 beside the interstate.

Home again, and Jane and I are going walkabout. I have her rigged on my shoulders in the backpack. Distributed throughout the aluminum frame and snugged straps, her weight dissipates to nothing. After all, she weighs little more than a good-sized chicken. As we step into the yard, I twist my neck to get a look at her face and find her looking out over the valley below. Her eyes are wide and steady beneath the brim of her floppy cap. How far out of infancy do we lose this gaze, with its utter absence of expectation or prejudice? What is it like to simply *see* what is before you, without the skew of context?

We begin on the easy path—a mown strip leading to the ridge past the old circular steel corncrib behind the granary. The crib stands empty beneath its rust-streaked galvanized cap, the iron mesh twined around the south side with a few stray ivy runners. For years it has done little more than sift the wind. At sundown it silhouettes against the sky like some ghostly aviary.

The leaves are well-turned and beginning to fall. Pale brown swatches of ripening corn stripe the far hillside, and crimson swatches of sumac fill the swales like coals banked against winter. The clouds are wispy in a pale blue sky, and the air is just crisp enough that you can imagine the smell of burning leaves despite the clear air. On the unpastured hillsides the tall grass is gone lank and set to fade.

We move off the path and ease downhill into the waist-high grasses. A few weeks ago and I would be stirring up a steady click and whir of fleeing insects—now there is just the occasional grasshopper and a smattering of small ground moths. With each step I'm knocking loose seeds and husks—several of them find their way into my socks. This is an interesting corner of the farm—old overgrown grassland pasture rounding off and rolling steeply into patchwork groves. They shelter a valley where centuries of spring runoff have cut two sharp draws that converge to run in a single ravine westward. The bottomland trees are gnarled and fat, and twisted in mysterious ways, and they grow overlooking sharp banks and sinuous trenches. One is so unusually configured with fat low-hanging limbs and knotholes that Amy has dubbed it her Magic Tree.

Here in the old pasture, there are a few young pine trees—all under six feet and planted by my mother-in-law and the owners previous to her—but mostly the open space is being taken over by box elders. Right at the tree line I come to the old barbed-wire fence. Much of it is still in decent shape—the galvanized wire loose from the posts here and there, and crushed by fallen trees in a couple of spots, but it hasn't gone rusty, and it wouldn't take much fixing. Anneliese and I have talked of grazing sheep out here, or getting some beef cows. Fixing this old fence and putting up new is on our wish list for next year. I follow the fence line for a while and find a couple of spots where the wire has been swallowed by the trees, grown deep inside the wood, and it hits me how much easier it is to speak of fixing fence than it actually will be to accomplish the task. I wonder too about clearing all those box elders, and if we'll have to fence the pine seedlings in if we hope for them to survive the cattle.

I cross the fence and go into the trees now, careful to hold the branches clear as I push through. Only ten feet into the canopy the feel of the place changes. Out in the field there was a sense of sweep and contour—in here with nothing but leaf scrap covering the ground between the big-trunked trees, I get that secret hide-out feeling, the same little tingle low in the gut that I got when Ricky and I would hide out in the canary grass along Beaver Creek Road. Down here among the big trees with the sky closed mostly out, things are a gray shade of brown, so when I spot a cluster of brilliant red berries it is like a gift, and I stop to study them, kneeling down and tipping forward so that Jane might see. I talk to her quietly, reveling in the joy of being out on the skin of this rough earth, heads in the cool atmosphere of infinity, and yet able to speak so quietly and be heard. Jane wraps both little fists around the aluminum frame and amuses herself by chewing the nylon. I hike back out into the open and upward, and when I reach the ridge she is still champing happily away. I can see her chubby little arm hanging over the edge of the pack, bouncing in time to the pace we are keeping. I put my hand back over my shoulder, palm up. I see the little hand reaching now, slowly, until she lays her teensy paw in mine, then clasps her fingers around my thumb, and I look to the blue sky and think a silent *Thank you.*

Anneliese and Amy have gone to the neighbors to get a pickup load of straw for mulch, and I get a little zoom as I always do when I see my wife driving the pickup truck. She sets to digging potatoes in the garden, and I head for the office. Amy is unloading the straw, and Jane is happily struggling to all fours in the garden dirt as the chickens scratch and peck around her. Maple leaves are petaling down, and Jane smashes one in her fist, then shoves it in her mouth. Anneliese is beautiful with a touch of color in her cheekbones, but she also looks tired. I like to make jokes and goof on my own incompetence, but the truth is, this year has stretched my wife beyond anything that is fair. I must find a better way to navigate. As much as I love the animals, I know where my bread and butter lies, and future adjustments may have to take that into consideration. All those times I told smart-aleck stories about farming, while back home my wife fed the pigs. Even more humiliating, next week a man will bring a load of firewood—all my selfish solo chopping, and still I didn't split enough for winter.

The backbeat of this year—and it's laid in there deep, you have to listen for it—is that I am trying to do too much, and I'm not the one paying for it. I haven't cooked a meal with my wife in months. The pantry is full with home canning, and I spent maybe four hours in the garden. The division of labor has become nigh unto no division at all. When my dad was milking all those cows, I still used to see him grab a broom and sweep the kitchen now and then. Lately Anneliese has been doing work as a freelance translator, and when I see her dressed up and leaving the house in a professional capacity I am simultaneously proud and ashamed that I may be depriving her of more of that. In short, I want to be a better husband and a better father, and the most meaningful progress in that direction requires me to do one simple thing: Be There; or better yet, Be Here.

This morning when I go out to feed the chickens, my boots leave a swipe of tracks through the frost. Soon I'll have to rig a deal to keep the chickens' water from freezing, and hang a lightbulb on a timer for the worst winter nights. The coop is still unpainted, and I have yet to nail up the trim boards Mills cut to fit the eaves. The structure itself is sitting solid, but just as Buffalo and I placed it, it remains tipped a good bit off plumb. One local wag refers to it as the Leaning Tower of Poultry. When I pull the door open—the door it took me six tries to get right on Mills's scorching blacktop that day—there are the six multicolored ladies, beadily blinking and ready for the day. I scoop fresh feed into a feeder fashioned from two scraps of plywood tacked in a vee between a pair of one-by-four boards (a rare carpentry triumph—I found the instructions in a library book) and replenish their water. While the chickens dip and peck I raid the nesting boxes. Dad built the boxes one day when I was feeling especially behind, and then he and Amy hung them.

There are three eggs this morning, two of them warm. Likely there will be one or two more by the afternoon, as lately we've been nearing consistently peak production. I drop one of the miniature doors open and the surviving Barred Rock pokes her head out first. When I look back just before going into the house, three of the hens are out, tilting their heads curiously at the frost.

And then it's out of the cold air and into the warm air of the kitchen, and the sound of bacon in the pan. I sure liked having those pigs around, but Bob the One-Eyed Beagle and his crew cure that bacon slow and smoke it with real wood, and at first whiff all residual reservations vaporize. Anneliese is frying potatoes and onions in a cast iron pan. I dice up some tomatoes and garlic, and while they sauté I whip the eggs for scrambling. Amy is setting the table, and Jane is burbling in her baby seat.

The ceiling fan in the living room is pushing the heat from the woodstove back to the floor and into the kitchen. We sit down in our small circle for a breakfast in which everything but the salt, pepper, and olive oil came to the table by our own hand. Here we are in the slantways house, the fire warm, our plates full, our chickens tiptoeing from their crooked coop out there on the hill.

EPILOGUE

Before our family grew large, my brother John and I shared the north bedroom of the farmhouse. The town road ran north-south past the garden, and at night the headlights of south-bound cars pushed through the windows and slid across the plaster. A shifting rectangle of light would appear on the wall beyond the foot of my bed, pass slowly to the right, then bend around the corner and work back toward the head end, all the while growing narrower and narrower until the rectangle squeezed to nothing and the room was dark again. Traffic was rare back then, and the slippery patch of light always got you wondering where people were going.

Every night, Dad would climb the bare wooden stairs with his Bible in hand. He'd seat himself on the edge of the bed and read a chapter aloud. I can still recall his weight on the mattress, the way it drew me toward him. He'd leaf a bit to find his place, the parchment pages all whisper and crinkle. There were a lot of books in our house. None of them sounded like that book.

He read the chapters in sequence, one per night. He read steadily, with neither adornment nor portent. Just the way, as a matter of fact, that he lived. At the end of the chapter, he rose, tweaked our covers, bid us good night, and left the room. He'd snap the light switch on his way out the door, and start in on a hymn. It is one of the bedrock memories of my childhood, him singing as his footsteps receded down the hall. I can picture his toed-out gait, accentuated by the Li'l Abner curl of his leather work boots. He sang the way he read, purely and plainly, although he had a tendency to hold back on a syllable now and then and drop it behind the beat, just a dab of jazz. His voice echoed up the stairwell until he was downstairs and the verse was done. Shortly we would hear the lullaby murmur of our parents in conversation, and the clink of a spoon on a bowl as Dad had his bedtime cereal or a dish of ice cream.

How warm we were in our beds, watching the light slip silent around the room until it shrank into darkness and we went heavy-lidded to sleep.

ACKNOWLEDGMENTS

First and foremost, to my parents—anything decent is because of them, anything else is not their fault.

Gene Logsdon, Ben Logan, and Jerry Apps—country chroniclers long before I tossed my first forkful.

John and Julie—ever since we got out of prison, things have been going great. Mills (redneck doula and amateur body piercer), Billy and Margie (Knuckles, R.I.P.), Buffalo (Gosh, I hope it's sunny for the next thirty years), Racy's and the Racy's crew for low lighting, late hours, drop-shipping, and a tab (and a special hello to Mister Happy for making me appear to be Miss Congeniality by comparison. Are those beans dolphin-safe?). Karen Rose for math. Krister for the usual rescues. Matt Marion for work and cat stories.

Robert Gough, from whose book my cutover synopsis was drawn. Wisconsin Historical Society, the curators of www.mon archrange.com, and Mary Beth Jacobson at the Dodge County Historical Museum.

Our Colorado family. ALR (giving public radio some low end). McDowells—indulging me now for decades.

Alison for the start, Jennifer for the finish (are we there yet?), Jeanette Perez, Rachel Elinsky, Jason Sack. Lisa, Tina, and now Elizabeth. Scranton (books in boxes!). Mags everytime.

Frank for both confirming and kicking the compass, plus the barber chair is *back*!

Blakeley—from Blue Hills to booking, what a road.

Alissa for handling it.

Everyone on the road and at the readings. My friend Ben.

Men's Health, No Depression, Wisconsin People & Ideas, Hope, Encyclopedia of the Midwest, Backpacker Magazine, Farm Life, the anthology *Seasons on the Farm* (in particular Lee Klancher and Amy Glaser), Wisconsin Public Television (specifically Banjo Boy and the Vet's Girl) and the Fond du Lac Public Library for publishing essays or producing projects from which some of the material for this book was drawn.

Our new Fall Creek neighbors. We're lucky to be here.

Donna for finding these acres.

Nobbern, the only place I'll ever be from. See you at the All-School Reunion. Here's yer chicken.

If I missed you, knock firmly, step clear, and keep your hands in sight. . . . I'll be right down, but not right out.

And in ways the world will and will not know, I love and thank the shaman girl, Snorticus the horse girl, and She Who is Still Emerging (but is so far trending LOUD).